AF810914

RILEM State-of-the-Art Reports

RILEM STATE-OF-THE-ART REPORTS
Volume 38

RILEM, The International Union of Laboratories and Experts in Construction Materials, Systems and Structures, founded in 1947, is a non-governmental scientific association whose goal is to contribute to progress in the construction sciences, techniques and industries, essentially by means of the communication it fosters between research and practice. RILEM's focus is on construction materials and their use in building and civil engineering structures, covering all phases of the building process from manufacture to use and recycling of materials. More information on RILEM and its previous publications can be found on www.RILEM.net.

The RILEM State-of-the-Art Reports (STAR) are produced by the Technical Committees. They represent one of the most important outputs that RILEM generates – high level scientific and engineering reports that provide cutting edge knowledge in a given field. The work of the TCs is one of RILEM's key functions.

Members of a TC are experts in their field and give their time freely to share their expertise. As a result, the broader scientific community benefits greatly from RILEM's activities.

RILEM's stated objective is to disseminate this information as widely as possible to the scientific community. RILEM therefore considers the STAR reports of its TCs as of highest importance, and encourages their publication whenever possible.

The information in this and similar reports is mostly pre-normative in the sense that it provides the underlying scientific fundamentals on which standards and codes of practice are based. Without such a solid scientific basis, construction practice will be less than efficient or economical.

It is RILEM's hope that this information will be of wide use to the scientific community.

Indexed in SCOPUS, Google Scholar and SpringerLink.

Augusto Cannone Falchetto · Lily Poulikakos ·
Emiliano Pasquini · Di Wang
Editors

Valorisation of Waste and Secondary Materials for Roads

State-of-the-Art Report of the RILEM TC 279-WMR

rilem

Springer

Editors
Augusto Cannone Falchetto
Department of Civil Engineering
Aalto University
Espoo, Finland

Lily Poulikakos
308—Concrete and Asphalt
EMPA
Dübendorf, Switzerland

Emiliano Pasquini
DICEA
University of Padua
Padova, Italy

Di Wang
Department of Civil Engineering
Aalto University
Espoo, Finland

ISSN 2213-204X ISSN 2213-2031 (electronic)
RILEM State-of-the-Art Reports
ISBN 978-3-031-33172-5 ISBN 978-3-031-33173-2 (eBook)
https://doi.org/10.1007/978-3-031-33173-2

© RILEM 2023

No part of this work may be reproduced, stored in a retrieval system, or transmitted in any form or by any means, electronic, mechanical, photocopying, microfilming, recording or otherwise, without written permission from the Publisher, with the exception of any material supplied specifically for the purpose of being entered and executed on a computer system, for exclusive use by the purchaser of the work. Permission for use must always be obtained from the owner of the copyright: RILEM.
The use of general descriptive names, registered names, trademarks, service marks, etc. in this publication does not imply, even in the absence of a specific statement, that such names are exempt from the relevant protective laws and regulations and therefore free for general use.
The publisher, the authors and the editors are safe to assume that the advice and information in this book are believed to be true and accurate at the date of publication. Neither the publisher nor the authors or the editors give a warranty, expressed or implied, with respect to the material contained herein or for any errors or omissions that may have been made. The publisher remains neutral with regard to jurisdictional claims in published maps and institutional affiliations.

This Springer imprint is published by the registered company Springer Nature Switzerland AG
The registered company address is: Gewerbestrasse 11, 6330 Cham, Switzerland

Preface

The position of TC 279-WMR in RILEM Cluster F: Bituminous Materials and Polymers

Investigating the possibility of incorporating waste, marginal and secondary materials in pavement has become a critical research area in the field of road engineering as, over the years, the sensibility and attention to the sustainability of road infrastructure have increased. At the same time, this concept of using diverse potentially valuable materials derived from different production activities and industrial by-products was considered of significant importance within the scientific community and the RILEM cluster F on Bituminous Materials and Polymers. TC 279-WMR was the first technical committee to propose, organize, and perform structured research on Waste and Secondary Materials for Roads. Being an entirely new committee, TC 279-WMR was able to approach the novel research field with sufficient freedom in selecting the topics and aspects to be addressed with the awareness of the complexity and challenges that might be experienced when exploring unknown or unfamiliar areas. Nevertheless, TC 279-WMR could significantly benefit from the research effort and output of the past and active RILEM technical committees. Experimental methods, chemo-rheological analysis, and recycling solutions are among the tools that were adopted from the following TCs: TC 231-NBM (Nano-Technology-Based Bituminous Materials), TC 237-SIB (Testing and Characterization of Sustainable Innovative Bituminous Materials and Systems), TC 241-MCD (Mechanics of Cracking in Asphalt and Composite Pavements), TC 252-CMB (Chemo-Mechanical Characterization of Bituminous Materials), TC 264-RAP (Recycling of Asphalt Pavements), and TC 272-PIM (Phase and Interphase behavior of innovative bituminous materials). The research outcome of TC 279-WMR represents the results of a seminal work and the basis for continuing the research on alternative materials for paving applications in follow-up RILEM activities with the aim of synergistic cooperation with other technical committees while promoting and widely disseminating the research output.

Goals of TC 279-WMR

Over the past five years, the RILEM Technical Committee 279-WMR planned, organized, and performed an extensive scientific investigation to deepen the comprehension and knowledge concerning the utilization of waste and marginal resources in materials for pavements by conducting interlaboratory research to address their behavior and preparing novel methodologies on the selection, use, and application thereof.

Under the supervision of the Chair, Dr. Lily Poulikakos, EMPA, Switzerland, and the Deputy Chair, Prof. Emiliano Pasquini, University of Padova, Italy, TC 279-WMR was structured into four task groups (TGs) as listed below [1]:

- TG1 *Bituminous binder and bituminous mixture modified with waste polyethylene* (TG leader: Marjan Tušar, Slovenian National Building and Civil Engineering Institute, Slovenia)
- TG2 *Crumb Rubber Modified Binders* (TG leader: Jorge C. Pais, University of Minho, Portugal)
- TG3 *Waste aggregates in asphalt mixtures such as construction and demolition waste, recycled concrete aggregates, and steel slag and cooperating with TG1 on the incorporation of polyethylene in asphalt mixtures* (TG leaders: Emiliano Pasquini, University of Padova, Italy; Augusto Cannone Falchetto, Aalto University, Finland; Fernando Moreno-Navarro, University of Granada, Spain)
- TG4 *Life Cycle Assessment of asphalt mixtures with WMR* (TG leaders: Davide Lo Presti, University of Palermo, Italy; Ana Jiménez del Barco Carrión, University of Granada, Spain)

The outputs of the technical committee were disseminated at different events. The initial results of the research were presented at the 1st RILEM International Symposium on Bituminous Materials (ISBM Lyon 2020), 9 December 17, 2020 [2]; the flyer is attached below. Two additional workshops will be organized during the international conferences of the 10th European Asphalt Technology Association (EATA 2023) (June 12–14, 2023, Gdansk, Poland) and the 13th International Conference on Road and Airfield Pavement Technology (ICPT 2023) (July 6–8, 2023, Beijing, China). The key TC findings will be presented during the workshops. Further dissemination will occur at the 77th RILEM Annual Week and the 1st Interdisciplinary Symposium on Smart and Sustainable Infrastructures (September 4–8, 2023, Vancouver, Canada). In addition, recommendation documents and the different aspects of using secondary materials for roads are under preparation and will be published soon after the present book.

This State-of-the-Art (STAR) Report synthesizes the work performed by TC 279-WMR, summarizing the research on the utilization of waste and marginal materials in asphalt pavements. The structure of the STAR Report follows the organization of the TC and its TGs and related areas of investigation (i.e., implementation of plastic in binder and mixture; application of crumb rubber in asphalt binder; use of construction demolition waste, concrete aggregate, and steel slags in asphalt mixture; and life cycle assessment).

Acknowledgments

The four editors of this STAR document would like to sincerely acknowledge colleagues, members, and friends of the TC 279-WMR who contributed in any form to the success of the TC over the years (2017-2022). The exchange of ideas, comments, and feedback during several meetings, conferences, and events enriched the scientific diversity of the committee, promoting the effort toward the excellence that RILEM activities demand. Special appreciation to the active participants that authored conference and journal publications supporting the dissemination of the TC activities in different venues. Thank you to all of you!

TC 279-WMR Members

Gordon Airey, University of Nottingham, UK
Martin Arraigada, Empa, Switzerland
Hassan Baaj, University of Waterloo, Canada
Liedi Bernucci, University of São Paulo, Brazil
Johan Blom, University of Antwerp, Belgium
Nicolas Bueche, Bern University of Applied Sciences, Switzerland
Moises Bueno, Empa, Switzerland; Sika Group, Switzerland
Augusto Cannone Falchetto, Aalto University, Finland
Xavier Carbonneau, Colas Group, France
Jean Claude Carret, École de technologie supérieure ÉTS, Montréal, Canada
Alan Carter, École de technologie supérieure ÉTS, Montréal, Canada
Maria Chiara Cavalli, KTH Royal Institute of Technology, Sweden
Emmanuel Chailleux, IFSTTAR, France
Davide Dalmazzo, Polytechnic University of Turin, Italy
Herve Di Benedetto, University of Lyon/ENTPE, France
Elham Fini, Arizona State University, USA
Anna Formisano, University of Naples Federico II, Italy
Ana Cristina Freire, National Laboratory for Civil Engineering, Portugal
Vincent Gaudefroy, IFSTTAR, France
Giovanni Giacomello, University of Padova, Italy
Andrea Graziani, Marche Polytechnic University, Italy
James Grenfell, Australian Road Research Board, Australia
Fucheng Guo, Aalto University, Finland
Ankit Gupta, Indian Institute of Technology, India
Guillaume Habert, ETH Zurich, Switzerland

Ailar Hajimohammadi, University of New South Wales, Australia
David Hernando, University of Antwerp, Belgiu
Bernhard Hofko, Vienna University of Technology, Austria
Ana Jimenez del Barco Carrion, University of Granada, Spain
Ruxin Jing, Delft University of Technology, Netherlands
Rouba Joumblat, Beirut Arab University, Lebanon
Muhammad Rafiq Khan Kakar, Bern University of Applied Sciences, Switzerland
Patricia Kara De Maeijer, University of Antwerp, Belgium
Alalea Kia, Imperial College London, UK
Greet Leegwater, Delft University of Technology/TNO, Netherlands
Tung-Chai (Bill) Ling, Hunan University, China
Davide Lo Presti, University of Palermo, Italy
Xiaohu Lu, Nynas AB, Sweden
Salvatore Mangiafico, University of Lyon/ENTPE, France
Oumaya Marzouk, Cerema Centre Est—Laboratoire d'AUTUN, France
Meng Guo, Beijing University of Technology, China
David Mensching, Federal Highway Administration, USA
Peter Mikhailenko, Empa, Switzerland
Miomir Miljković, University of Niš, Serbia
Ben Moins, University of Antwerp, Belgium
Konrad Mollenhauer, University of Kassel, Germany
Fernando Moreno-Navarro, University of Granada, Spain
Jose Norambuena, University of the Bío Bío, Chile
Marko Oreskovic, University of Belgrade, Serbia
Gabriel Orozco, École de technologie supérieure ÉTS, Montréal, Canada
Jorge Carvalho Pais, University of Minho, Portugal
Manfred N. Partl, Empa, Switzerland
Marco Pasetto, University of Padova, Italy
Emiliano Pasquini, University of Padova, Italy
Daniel Perraton, École de technologie supérieure ÉTS, Montréal, Canada
Zhengyin Piao, Empa, Switzerland; Yale University, USA
Laurent Porot, Kraton Polymers B.V., Netherlands
Lily Poulikakos, Empa, Switzerland
Jian Qiu, AsfaltNu C.V., Netherlands
Simone Raschia, STS s.r.l., Italy
Chiara Riccardi, Pisa University, Italy
Mayca Rubio Gamez, University of Granada, Spain
Nikhil Saboo, Indian Institute of Technology, India
Cesare Sangiorgi, University of Bologna, Italy
Cedric Sauzeat, University of Lyon/ENTPE, France
Mohammad Shafiee, National Research Council-CNRC, Canada
Marta Skaf, University of Burgos, Spain
Judita Škulteckė, Vilnius Gediminas Technical University, Lithuania
Marjan Tušar, Slovenian National Building and Civil Engineering Institute, Slovenia
Michel Vaillancourt, École de technologie supérieure ÉTS, Montréal, Canada

Wim Van den bergh, University of Antwerp, Belgium
Katerina Varveri, Delft University of Technology, Netherlands
Kamila Vasconcelos, University of São Paulo, Brazil
Rosa Veropalumbo, University of Naples Federico II, Italy
Nunzio Viscione, University of Naples Federico II, Italy
Giovanni Volpatti, Cemex, Switzerland
Di Wang, Aalto University, Finland
Haopeng Wang, University of Nottingham, UK
Michael Wistuba, Technical University of Braunschweig, Germany
Martins Zaumanis, Empa, Switzerland
Adam Zofka, FRUIT Foundation, Poland

Non RILEM Member Contributors

Hessam Azari Jafari, Massachusetts Institute of Technology, USA
Andrea Baliello, University of Padova, Italy
Ana Benavent, Valoriza Servicios Medioambientales, SA, Spain
Ramon Botella, Universitat Politècnica de Catalunya·Barcelona Tech, Spain
Gabriella Buttitta, University of Palermo, Italy
Gaetano Di Mino, University of Palermo, Italy
Abdeljebbar Diouri, Mohammed V University, Morocco
Michael Elwardany, Florida State University, USA
Marie Enfrin, Royal Melbourne Institute of Technology University, Australia
Wellington Lorran Gaia Ferreira, Federal Rural University of the Semi-arid Region, Brazil
Gaspare Giancontieri, University of Palermo, Italy
Meng Guo, Beijing University of Technology, China
Biruk Hailesilassie, Kistler Instrumente AG, Switzerland
Norbert Heeb, Empa, Switzerland
Agarwal Himanshu, Zydex Industries, India
Atsushi Kawakami, Public Works Research Institute, Japan
Jorge Manuel Barreira De Sousa, Consulpav, Portugal
Sonia Megert, Tyre Recycling Solutions, Switzerland
Guillaume Rousseau, Colas, France
Francesca Russo, University of Naples Federico II, Italy
Nicolas Schüwer, Tyre Recycling Solutions, Switzerland
Mingjiang Tao, Worcester Polytechnic Institute, USA
Raul Tauste-Martinez, University of Granada, Spain
Mªdela O Terciado, Valoriza Servicios Medioambientales, SA, Spain
Castelo Branco Verônica, Universidade Federal do Ceará, Brazil
Julien Waligora, Eiffage Infrastructures, France

Special thanks to the TG leaders and the authors of the chapters of this State-of-the-Art Report.

Espoo, Finland Augusto Cannone Falchetto
Dübendorf, Switzerland Lily Poulikakos
Padova, Italy Emiliano Pasquini
Espoo, Finland Di Wang

References

RILEM TC 279-WMR Homepage, https://www.rilem.net/groupe/279-wmr-valorisation-of-waste-and-secondary-materials-for-roads-369, last accessed 2022/12/03
ISBM Homepage, https://isbmlyon2020.sciencesconf.org, last accessed 2022/10/21

RILEM Publications

The following list is presenting the global offer of RILEM Publications, sorted by series. Each publication is available in printed version and/or in online version.

RILEM Proceedings (PRO)

PRO 1: Durability of High Performance Concrete (ISBN: 2-912143-03-9; e-ISBN: 2-351580-12-5; e-ISBN: 2351580125); *Ed. H. Sommer*

PRO 2: Chloride Penetration into Concrete (ISBN: 2-912143-00-04; e-ISBN: 2912143454); *Eds. L.-O. Nilsson and J.-P. Ollivier*

PRO 3: Evaluation and Strengthening of Existing Masonry Structures (ISBN: 2-912143-02-0; e-ISBN: 2351580141); *Eds. L. Binda and C. Modena*

PRO 4: Concrete: From Material to Structure (ISBN: 2-912143-04-7; e-ISBN: 2351580206); *Eds. J.-P. Bournazel and Y. Malier*

PRO 5: The Role of Admixtures in High Performance Concrete (ISBN: 2-912143-05-5; e-ISBN: 2351580214); *Eds. J. G. Cabrera and R. Rivera-Villarreal*

PRO 6: High Performance Fiber Reinforced Cement Composites—HPFRCC 3 (ISBN: 2-912143-06-3; e-ISBN: 2351580222); *Eds. H. W. Reinhardt and A. E. Naaman*

PRO 7: 1st International RILEM Symposium on Self-Compacting Concrete (ISBN: 2-912143-09-8; e-ISBN: 2912143721); *Eds. Å. Skarendahl and Ö. Petersson*

PRO 8: International RILEM Symposium on Timber Engineering (ISBN: 2-912143-10-1; e-ISBN: 2351580230); *Ed. L. Boström*

PRO 9: 2nd International RILEM Symposium on Adhesion between Polymers and Concrete ISAP '99 (ISBN: 2-912143-11-X; e-ISBN: 2351580249); *Eds. Y. Ohama and M. Puterman*

PRO 10: 3rd International RILEM Symposium on Durability of Building and Construction Sealants (ISBN: 2-912143-13-6; e-ISBN: 2351580257); *Ed. A. T. Wolf*

PRO 11: 4th International RILEM Conference on Reflective Cracking in Pavements (ISBN: 2-912143-14-4; e-ISBN: 2351580265); *Eds. A. O. Abd El Halim, D. A. Taylor and El H. H. Mohamed*

PRO 12: International RILEM Workshop on Historic Mortars: Characteristics and Tests (ISBN: 2-912143-15-2; e-ISBN: 2351580273); *Eds. P. Bartos, C. Groot and J. J. Hughes*

PRO 13: 2nd International RILEM Symposium on Hydration and Setting (ISBN: 2-912143-16-0; e-ISBN: 2351580281); *Ed. A. Nonat*

PRO 14: Integrated Life-Cycle Design of Materials and Structures—ILCDES 2000 (ISBN: 951-758-408-3; e-ISBN: 235158029X); (ISSN: 0356-9403); *Ed. S. Sarja*

PRO 15: Fifth RILEM Symposium on Fibre-Reinforced Concretes (FRC)—BEFIB'2000 (ISBN: 2-912143-18-7; e-ISBN: 291214373X); *Eds. P. Rossi and G. Chanvillard*

PRO 16: Life Prediction and Management of Concrete Structures (ISBN: 2-912143-19-5; e-ISBN: 2351580303); *Ed. D. Naus*

PRO 17: Shrinkage of Concrete—Shrinkage 2000 (ISBN: 2-912143-20-9; e-ISBN: 2351580311); *Eds. V. Baroghel-Bouny and P.-C. Aïtcin*

PRO 18: Measurement and Interpretation of the On-Site Corrosion Rate (ISBN: 2-912143-21-7; e-ISBN: 235158032X); *Eds. C. Andrade, C. Alonso, J. Fullea, J. Polimon and J. Rodriguez*

PRO 19: Testing and Modelling the Chloride Ingress into Concrete (ISBN: 2-912143-22-5; e-ISBN: 2351580338); *Eds. C. Andrade and J. Kropp*

PRO 20: 1st International RILEM Workshop on Microbial Impacts on Building Materials (CD 02) (e-ISBN 978-2-35158-013-4); *Ed. M. Ribas Silva*

PRO 21: International RILEM Symposium on Connections between Steel and Concrete (ISBN: 2-912143-25-X; e-ISBN: 2351580346); *Ed. R. Eligehausen*

PRO 22: International RILEM Symposium on Joints in Timber Structures (ISBN: 2-912143-28-4; e-ISBN: 2351580354); *Eds. S. Aicher and H.-W. Reinhardt*

PRO 23: International RILEM Conference on Early Age Cracking in Cementitious Systems (ISBN: 2-912143-29-2; e-ISBN: 2351580362); *Eds. K. Kovler and A. Bentur*

PRO 24: 2nd International RILEM Workshop on Frost Resistance of Concrete (ISBN: 2-912143-30-6; e-ISBN: 2351580370); *Eds. M. J. Setzer, R. Auberg and H.-J. Keck*

PRO 25: International RILEM Workshop on Frost Damage in Concrete (ISBN: 2-912143-31-4; e-ISBN: 2351580389); *Eds. D. J. Janssen, M. J. Setzer and M. B. Snyder*

PRO 26: International RILEM Workshop on On-Site Control and Evaluation of Masonry Structures (ISBN: 2-912143-34-9; e-ISBN: 2351580141); *Eds. L. Binda and R. C. de Vekey*

PRO 27: International RILEM Symposium on Building Joint Sealants (CD03; e-ISBN: 235158015X); *Ed. A. T. Wolf*

PRO 28: 6th International RILEM Symposium on Performance Testing and Evaluation of Bituminous Materials—PTEBM'03 (ISBN: 2-912143-35-7; e-ISBN: 978-2-912143-77-8); *Ed. M. N. Partl*

PRO 29: 2nd International RILEM Workshop on Life Prediction and Ageing Management of Concrete Structures (ISBN: 2-912143-36-5; e-ISBN: 2912143780); *Ed. D. J. Naus*

PRO 30: 4th International RILEM Workshop on High Performance Fiber Reinforced Cement Composites—HPFRCC 4 (ISBN: 2-912143-37-3; e-ISBN: 2912143799); *Eds. A. E. Naaman and H. W. Reinhardt*

PRO 31: International RILEM Workshop on Test and Design Methods for Steel Fibre Reinforced Concrete: Background and Experiences (ISBN: 2-912143-38-1; e-ISBN: 2351580168); *Eds. B. Schnütgen and L. Vandewalle*

PRO 32: International Conference on Advances in Concrete and Structures 2 vol. (ISBN (set): 2-912143-41-1; e-ISBN: 2351580176); *Eds. Ying-shu Yuan, Surendra P. Shah and Heng-lin Lü*

PRO 33: 3rd International Symposium on Self-Compacting Concrete (ISBN: 2-912143-42-X; e-ISBN: 2912143713); *Eds. Ó. Wallevik and I. Níelsson*

PRO 34: International RILEM Conference on Microbial Impact on Building Materials (ISBN: 2-912143-43-8; e-ISBN: 2351580184); *Ed. M. Ribas Silva*

PRO 35: International RILEM TC 186-ISA on Internal Sulfate Attack and Delayed Ettringite Formation (ISBN: 2-912143-44-6; e-ISBN: 2912143802); *Eds. K. Scrivener and J. Skalny*

PRO 36: International RILEM Symposium on Concrete Science and Engineering—A Tribute to Arnon Bentur (ISBN: 2-912143-46-2; e-ISBN: 2912143586); *Eds. K. Kovler, J. Marchand, S. Mindess and J. Weiss*

PRO 37: 5th International RILEM Conference on Cracking in Pavements—Mitigation, Risk Assessment and Prevention (ISBN: 2-912143-47-0; e-ISBN: 2912143764); *Eds. C. Petit, I. Al-Qadi and A. Millien*

PRO 38: 3rd International RILEM Workshop on Testing and Modelling the Chloride Ingress into Concrete (ISBN: 2-912143-48-9; e-ISBN: 2912143578); *Eds. C. Andrade and J. Kropp*

PRO 39: 6th International RILEM Symposium on Fibre-Reinforced Concretes—BEFIB 2004 (ISBN: 2-912143-51-9; e-ISBN: 2912143748); *Eds. M. Di Prisco, R. Felicetti and G. A. Plizzari*

PRO 40: International RILEM Conference on the Use of Recycled Materials in Buildings and Structures (ISBN: 2-912143-52-7; e-ISBN: 2912143756); *Eds. E. Vázquez, Ch. F. Hendriks and G. M. T. Janssen*

PRO 41: RILEM International Symposium on Environment-Conscious Materials and Systems for Sustainable Development (ISBN: 2-912143-55-1; e-ISBN: 2912143640); *Eds. N. Kashino and Y. Ohama*

PRO 42: SCC'2005—China: 1st International Symposium on Design, Performance and Use of Self-Consolidating Concrete (ISBN: 2-912143-61-6; e-ISBN: 2912143624); *Eds. Zhiwu Yu, Caijun Shi, Kamal Henri Khayat and Youjun Xie*

PRO 43: International RILEM Workshop on Bonded Concrete Overlays (e-ISBN: 2-912143-83-7); *Eds. J. L. Granju and J. Silfwerbrand*

PRO 44: 2nd International RILEM Workshop on Microbial Impacts on Building Materials (CD11) (e-ISBN: 2-912143-84-5); *Ed. M. Ribas Silva*

PRO 45: 2nd International Symposium on Nanotechnology in Construction, Bilbao (ISBN: 2-912143-87-X; e-ISBN: 2912143888); *Eds. Peter J. M. Bartos, Yolanda de Miguel and Antonio Porro*

PRO 46: Concrete Life'06—International RILEM-JCI Seminar on Concrete Durability and Service Life Planning: Curing, Crack Control, Performance in Harsh Environments (ISBN: 2-912143-89-6; e-ISBN: 291214390X); *Ed. K. Kovler*

PRO 47: International RILEM Workshop on Performance Based Evaluation and Indicators for Concrete Durability (ISBN: 978-2-912143-95-2; e-ISBN: 9782912143969); *Eds. V. Baroghel-Bouny, C. Andrade, R. Torrent and K. Scrivener*

PRO 48: 1st International RILEM Symposium on Advances in Concrete through Science and Engineering (e-ISBN: 2-912143-92-6); *Eds. J. Weiss, K. Kovler, J. Marchand, and S. Mindess*

PRO 49: International RILEM Workshop on High Performance Fiber Reinforced Cementitious Composites in Structural Applications (ISBN: 2-912143-93-4; e-ISBN: 2912143942); *Eds. G. Fischer and V. C. Li*

PRO 50: 1st International RILEM Symposium on Textile Reinforced Concrete (ISBN: 2-912143-97-7; e-ISBN: 2351580087); *Eds. Josef Hegger, Wolfgang Brameshuber and Norbert Will*

PRO 51: 2nd International Symposium on Advances in Concrete through Science and Engineering (ISBN: 2-35158-003-6; e-ISBN: 2-35158-002-8); *Eds. J. Marchand, B. Bissonnette, R. Gagné, M. Jolin and F. Paradis*

PRO 52: Volume Changes of Hardening Concrete: Testing and Mitigation (ISBN: 2-35158-004-4; e-ISBN: 2-35158-005-2); *Eds. O. M. Jensen, P. Lura and K. Kovler*

PRO 53: High Performance Fiber Reinforced Cement Composites—HPFRCC5 (ISBN: 978-2-35158-046-2; e-ISBN: 978-2-35158-089-9); *Eds. H. W. Reinhardt and A. E. Naaman*

PRO 54: 5th International RILEM Symposium on Self-Compacting Concrete (ISBN: 978-2-35158-047-9; e-ISBN: 978-2-35158-088-2); *Eds. G. De Schutter and V. Boel*

PRO 55: International RILEM Symposium Photocatalysis, Environment and Construction Materials (ISBN: 978-2-35158-056-1; e-ISBN: 978-2-35158-057-8); *Eds. P. Baglioni and L. Cassar*

PRO 56: International RILEM Workshop on Integral Service Life Modelling of Concrete Structures (ISBN 978-2-35158-058-5; e-ISBN: 978-2-35158-090-5); *Eds. R. M. Ferreira, J. Gulikers and C. Andrade*

PRO 57: RILEM Workshop on Performance of cement-based materials in aggressive aqueous environments (e-ISBN: 978-2-35158-059-2); *Ed. N. De Belie*

PRO 58: International RILEM Symposium on Concrete Modelling—CONMOD'08 (ISBN: 978-2-35158-060-8; e-ISBN: 978-2-35158-076-9); *Eds. E. Schlangen and G. De Schutter*

PRO 59: International RILEM Conference on On Site Assessment of Concrete, Masonry and Timber Structures—SACoMaTiS 2008 (ISBN set: 978-2-35158-061-5; e-ISBN: 978-2-35158-075-2); *Eds. L. Binda, M. di Prisco and R. Felicetti*

PRO 60: Seventh RILEM International Symposium on Fibre Reinforced Concrete: Design and Applications—BEFIB 2008 (ISBN: 978-2-35158-064-6; e-ISBN: 978-2-35158-086-8); *Ed. R. Gettu*

PRO 61: 1st International Conference on Microstructure Related Durability of Cementitious Composites 2 vol., (ISBN: 978-2-35158-065-3; e-ISBN: 978-2-35158-084-4); *Eds. W. Sun, K. van Breugel, C. Miao, G. Ye and H. Chen*

PRO 62: NSF/ RILEM Workshop: In-situ Evaluation of Historic Wood and Masonry Structures (e-ISBN: 978-2-35158-068-4); *Eds. B. Kasal, R. Anthony and M. Drdácký*

PRO 63: Concrete in Aggressive Aqueous Environments: Performance, Testing and Modelling, 2 vol., (ISBN: 978-2-35158-071-4; e-ISBN: 978-2-35158-082-0); *Eds. M. G. Alexander and A. Bertron*

PRO 64: Long Term Performance of Cementitious Barriers and Reinforced Concrete in Nuclear Power Plants and Waste Management—NUCPERF 2009 (ISBN: 978-2-35158-072-1; e-ISBN: 978-2-35158-087-5); *Eds. V. L'Hostis, R. Gens and C. Gallé*

PRO 65: Design Performance and Use of Self-consolidating Concrete—SCC'2009 (ISBN: 978-2-35158-073-8; e-ISBN: 978-2-35158-093-6); *Eds. C. Shi, Z. Yu, K. H. Khayat and P. Yan*

PRO 66: 2nd International RILEM Workshop on Concrete Durability and Service Life Planning—ConcreteLife'09 (ISBN: 978-2-35158-074-5; ISBN: 978-2-35158-074-5); *Ed. K. Kovler*

PRO 67: Repairs Mortars for Historic Masonry (e-ISBN: 978-2-35158-083-7); *Ed. C. Groot*

PRO 68: Proceedings of the 3rd International RILEM Symposium on 'Rheology of Cement Suspensions such as Fresh Concrete (ISBN 978-2-35158-091-2; e-ISBN: 978-2-35158-092-9); *Eds. O. H. Wallevik, S. Kubens and S. Oesterheld*

PRO 69: 3rd International PhD Student Workshop on 'Modelling the Durability of Reinforced Concrete (ISBN: 978-2-35158-095-0); *Eds. R. M. Ferreira, J. Gulikers and C. Andrade*

PRO 70: 2nd International Conference on 'Service Life Design for Infrastructure' (ISBN set: 978-2-35158-096-7, e-ISBN: 978-2-35158-097-4); *Eds. K. van Breugel, G. Ye and Y. Yuan*

PRO 71: Advances in Civil Engineering Materials—The 50-year Teaching Anniversary of Prof. Sun Wei' (ISBN: 978-2-35158-098-1; e-ISBN: 978-2-35158-099-8); *Eds. C. Miao, G. Ye and H. Chen*

PRO 72: First International Conference on 'Advances in Chemically-Activated Materials—CAM'2010' (2010), 264 pp., ISBN: 978-2-35158-101-8; e-ISBN: 978-2-35158-115-5; *Eds. Caijun Shi and Xiaodong Shen*

PRO 73: 2nd International Conference on 'Waste Engineering and Management—ICWEM 2010' (2010), 894 pp., ISBN: 978-2-35158-102-5; e-ISBN: 978-2-35158-103-2, *Eds. J. Zh. Xiao, Y. Zhang, M. S. Cheung and R. Chu*

PRO 74: International RILEM Conference on 'Use of Superabsorbent Polymers and Other New Additives in Concrete' (2010) 374 pp., ISBN: 978-2-35158-104-9; e-ISBN: 978-2-35158-105-6; *Eds. O.M. Jensen, M.T. Hasholt, and S. Laustsen*

PRO 75: International Conference on 'Material Science—2nd ICTRC—Textile Reinforced Concrete—Theme 1' (2010) 436 pp., ISBN: 978-2-35158-106-3; e-ISBN: 978-2-35158-107-0; *Ed. W. Brameshuber*

PRO 76: International Conference on 'Material Science—HetMat—Modelling of Heterogeneous Materials—Theme 2' (2010) 255 pp., ISBN: 978-2-35158-108-7; e-ISBN: 978-2-35158-109-4; *Ed. W. Brameshuber*

PRO 77: International Conference on 'Material Science—AdIPoC—Additions Improving Properties of Concrete—Theme 3' (2010) 459 pp., ISBN: 978-2-35158-110-0; e-ISBN: 978-2-35158-111-7; *Ed. W. Brameshuber*

PRO 78: 2nd Historic Mortars Conference and RILEM TC 203-RHM Final Workshop—HMC2010 (2010) 1416 pp., e-ISBN: 978-2-35158-112-4; *Eds. J. Válek, C. Groot and J. J. Hughes*

PRO 79: International RILEM Conference on Advances in Construction Materials Through Science and Engineering (2011) 213 pp., ISBN: 978-2-35158-116-2, e-ISBN: 978-2-35158-117-9; *Eds. Christopher Leung and K.T. Wan*

PRO 80: 2nd International RILEM Conference on Concrete Spalling due to Fire Exposure (2011) 453 pp., ISBN: 978-2-35158-118-6; e-ISBN: 978-2-35158-119-3; *Eds. E.A.B. Koenders and F. Dehn*

PRO 81: 2nd International RILEM Conference on Strain Hardening Cementitious Composites (SHCC2-Rio) (2011) 451 pp., ISBN: 978-2-35158-120-9; e-ISBN: 978-2-35158-121-6; *Eds. R.D. Toledo Filho, F.A. Silva, E.A.B. Koenders and E.M.R. Fairbairn*

PRO 82: 2nd International RILEM Conference on Progress of Recycling in the Built Environment (2011) 507 pp., e-ISBN: 978-2-35158-122-3; *Eds. V.M. John, E. Vazquez, S.C. Angulo and C. Ulsen*

PRO 83: 2nd International Conference on Microstructural-related Durability of Cementitious Composites (2012) 250 pp., ISBN: 978-2-35158-129-2; e-ISBN: 978-2-35158 123-0; *Eds. G. Ye, K. van Breugel, W. Sun and C. Miao*

PRO 84: CONSEC13—Seventh International Conference on Concrete under Severe Conditions—Environment and Loading (2013) 1930 pp., ISBN: 978-2-35158-124-7; e-ISBN: 978-2- 35158-134-6; *Eds. Z.J. Li, W. Sun, C.W. Miao, K. Sakai, O.E. Gjorv and N. Banthia*

PRO 85: RILEM-JCI International Workshop on Crack Control of Mass Concrete and Related issues concerning Early-Age of Concrete Structures—ConCrack 3—Control of Cracking in Concrete Structures 3 (2012) 237 pp., ISBN: 978-2-35158-125-4; e-ISBN: 978-2-35158-126-1; *Eds. F. Toutlemonde and J.-M. Torrenti*

PRO 86: International Symposium on Life Cycle Assessment and Construction (2012) 414 pp., ISBN: 978-2-35158-127-8, e-ISBN: 978-2-35158-128-5; *Eds. A. Ventura and C. de la Roche*

PRO 87: UHPFRC 2013—RILEM-fib-AFGC International Symposium on Ultra-High Performance Fibre-Reinforced Concrete (2013), ISBN: 978-2-35158-130-8, e-ISBN: 978-2-35158-131-5; *Eds. F. Toutlemonde*

PRO 88: 8th RILEM International Symposium on Fibre Reinforced Concrete (2012) 344 pp., ISBN: 978-2-35158-132-2; e-ISBN: 978-2-35158-133-9; *Eds. Joaquim A.O. Barros*

PRO 89: RILEM International workshop on performance-based specification and control of concrete durability (2014) 678 pp., ISBN: 978-2-35158-135-3; e-ISBN: 978-2-35158-136-0; *Eds. D. Bjegović, H. Beushausen and M. Serdar*

PRO 90: 7th RILEM International Conference on Self-Compacting Concrete and of the 1st RILEM International Conference on Rheology and Processing of Construction Materials (2013) 396 pp., ISBN: 978-2-35158-137-7; e-ISBN: 978-2-35158-138-4; *Eds. Nicolas Roussel and Hela Bessaies-Bey*

PRO 91: CONMOD 2014—RILEM International Symposium on Concrete Modelling (2014), ISBN: 978-2-35158-139-1; e-ISBN: 978-2-35158-140-7; *Eds. Kefei Li, Peiyu Yan and Rongwei Yang*

PRO 92: CAM 2014—2nd International Conference on advances in chemically-activated materials (2014) 392 pp., ISBN: 978-2-35158-141-4; e-ISBN: 978-2-35158-142-1; *Eds. Caijun Shi and Xiadong Shen*

PRO 93: SCC 2014—3rd International Symposium on Design, Performance and Use of Self-Consolidating Concrete (2014) 438 pp., ISBN: 978-2-35158-143-8; e-ISBN: 978-2-35158-144-5; *Eds. Caijun Shi, Zhihua Ou and Kamal H. Khayat*

PRO 94 (online version): HPFRCC-7—7th RILEM conference on High performance fiber reinforced cement composites (2015), e-ISBN: 978-2-35158-146-9; *Eds. H.W. Reinhardt, G.J. Parra-Montesinos and H. Garrecht*

PRO 95: International RILEM Conference on Application of superabsorbent polymers and other new admixtures in concrete construction (2014), ISBN: 978-2-35158-147-6; e-ISBN: 978-2-35158-148-3; *Eds. Viktor Mechtcherine and Christof Schroefl*

PRO 96 (online version): XIII DBMC: XIII International Conference on Durability of Building Materials and Components (2015), e-ISBN: 978-2-35158-149-0; *Eds. M. Quattrone and V.M. John*

PRO 97: SHCC3—3rd International RILEM Conference on Strain Hardening Cementitious Composites (2014), ISBN: 978-2-35158-150-6; e-ISBN: 978-2-35158-151-3; *Eds. E. Schlangen, M.G. Sierra Beltran, M. Lukovic and G. Ye*

PRO 98: FERRO-11—11th International Symposium on Ferrocement and 3rd ICTRC—International Conference on Textile Reinforced Concrete (2015), ISBN: 978-2-35158-152-0; e-ISBN: 978-2-35158-153-7; *Ed. W. Brameshuber*

PRO 99 (online version): ICBBM 2015—1st International Conference on Bio-Based Building Materials (2015), e-ISBN: 978-2-35158-154-4; *Eds. S. Amziane and M. Sonebi*

PRO 100: SCC16—RILEM Self-Consolidating Concrete Conference (2016), ISBN: 978-2-35158-156-8; e-ISBN: 978-2-35158-157-5; *Ed. Kamal H. Kayat*

PRO 101 (online version): III Progress of Recycling in the Built Environment (2015), e-ISBN: 978-2-35158-158-2; *Eds. I. Martins, C. Ulsen and S. C. Angulo*

PRO 102 (online version): RILEM Conference on Microorganisms-Cementitious Materials Interactions (2016), e-ISBN: 978-2-35158-160-5; *Eds. Alexandra Bertron, Henk Jonkers and Virginie Wiktor*

PRO 103 (online version): ACESC'16—Advances in Civil Engineering and Sustainable Construction (2016), e-ISBN: 978-2-35158-161-2; *Eds. T.Ch. Madhavi, G. Prabhakar, Santhosh Ram and P.M. Rameshwaran*

PRO 104 (online version): SSCS'2015—Numerical Modeling—Strategies for Sustainable Concrete Structures (2015), e-ISBN: 978-2-35158-162-9

PRO 105: 1st International Conference on UHPC Materials and Structures (2016), ISBN: 978-2-35158-164-3; e-ISBN: 978-2-35158-165 0

PRO 106: AFGC-ACI-fib-RILEM International Conference on Ultra-High-Performance Fibre-Reinforced Concrete—UHPFRC 2017 (2017), ISBN: 978-2-35158-166-7; e-ISBN: 978-2-35158-167-4; *Eds. François Toutlemonde and Jacques Resplendino*

PRO 107 (online version): XIV DBMC—14th International Conference on Durability of Building Materials and Components (2017), e-ISBN: 978-2-35158-159-9; *Eds. Geert De Schutter, Nele De Belie, Arnold Janssens and Nathan Van Den Bossche*

PRO 108: MSSCE 2016—Innovation of Teaching in Materials and Structures (2016), ISBN: 978-2-35158-178-0; e-ISBN: 978-2-35158-179-7; *Ed. Per Goltermann*

PRO 109 (2 volumes): MSSCE 2016—Service Life of Cement-Based Materials and Structures (2016), ISBN Vol. 1: 978-2-35158-170-4; Vol. 2: 978-2-35158-171-4; Set Vol. 1&2: 978-2-35158-172-8; e-ISBN : 978-2-35158-173-5; *Eds. Miguel Azenha, Ivan Gabrijel, Dirk Schlicke, Terje Kanstad and Ole Mejlhede Jensen*

PRO 110: MSSCE 2016—Historical Masonry (2016), ISBN: 978-2-35158-178-0; e-ISBN: 978-2-35158-179-7; *Eds. Inge Rörig-Dalgaard and Ioannis Ioannou*

PRO 111: MSSCE 2016—Electrochemistry in Civil Engineering (2016); ISBN: 978-2-35158-176-6; e-ISBN: 978-2-35158-177-3; *Ed. Lisbeth M. Ottosen*

PRO 112: MSSCE 2016—Moisture in Materials and Structures (2016), ISBN: 978-2-35158-178-0; e-ISBN: 978-2-35158-179-7; *Eds. Kurt Kielsgaard Hansen, Carsten Rode and Lars-Olof Nilsson*

PRO 113: MSSCE 2016—Concrete with Supplementary Cementitious Materials (2016), ISBN: 978-2-35158-178-0; e-ISBN: 978-2-35158-179-7; *Eds. Ole Mejlhede Jensen, Konstantin Kovler and Nele De Belie*

PRO 114: MSSCE 2016—Frost Action in Concrete (2016), ISBN: 978-2-35158-182-7; e-ISBN: 978-2-35158-183-4; *Eds. Marianne Tange Hasholt, Katja Fridh and R. Doug Hooton*

PRO 115: MSSCE 2016—Fresh Concrete (2016), ISBN: 978-2-35158-184-1; e-ISBN: 978-2-35158-185-8; *Eds. Lars N. Thrane, Claus Pade, Oldrich Svec and Nicolas Roussel*

PRO 116: BEFIB 2016—9th RILEM International Symposium on Fiber Reinforced Concrete (2016), ISBN: 978-2-35158-187-2; e-ISBN: 978-2-35158-186-5; *Eds. N. Banthia, M. di Prisco and S. Soleimani-Dashtaki*

PRO 117: 3rd International RILEM Conference on Microstructure Related Durability of Cementitious Composites (2016), ISBN: 978-2-35158-188-9; e-ISBN: 978-2-35158-189-6; *Eds. Changwen Miao, Wei Sun, Jiaping Liu, Huisu Chen, Guang Ye and Klaas van Breugel*

PRO 118 (4 volumes): International Conference on Advances in Construction Materials and Systems (2017), ISBN Set: 978-2-35158-190-2; Vol. 1: 978-2-35158-193-3; Vol. 2: 978-2-35158-194-0; Vol. 3: ISBN:978-2-35158-195-7; Vol. 4: ISBN:978-2-35158-196-4; e-ISBN: 978-2-35158-191-9; *Eds. Manu Santhanam, Ravindra Gettu, Radhakrishna G. Pillai and Sunitha K. Nayar*

PRO 119 (online version): ICBBM 2017—Second International RILEM Conference on Bio-based Building Materials, (2017), e-ISBN: 978-2-35158-192-6; *Eds. Sofiane Amziane and Mohammed Sonebi*

PRO 120 (2 volumes): EAC-02—2nd International RILEM/COST Conference on Early Age Cracking and Serviceability in Cement-based Materials and Structures, (2017), Vol. 1: 978-2-35158-199-5, Vol. 2: 978-2-35158-200-8, Set: 978-2-35158-197-1, e-ISBN: 978-2-35158-198-8; *Eds. Stéphanie Staquet and Dimitrios Aggelis*

PRO 121 (2 volumes): SynerCrete18: Interdisciplinary Approaches for Cement-based Materials and Structural Concrete: Synergizing Expertise and Bridging Scales of Space and Time, (2018), Set: 978-2-35158-202-2, Vol.1: 978-2-35158-211-4, Vol.2: 978-2-35158-212-1, e-ISBN: 978-2-35158-203-9; *Eds. Miguel Azenha, Dirk Schlicke, Farid Benboudjema, Agnieszka Knoppik*

PRO 122: SCC'2018 China—Fourth International Symposium on Design, Performance and Use of Self-Consolidating Concrete, (2018), ISBN: 978-2-35158-204-6, e-ISBN: 978-2-35158-205-3; *Eds. C. Shi, Z. Zhang, K. H. Khayat*

PRO 123: Final Conference of RILEM TC 253-MCI: Microorganisms-Cementitious Materials Interactions (2018), Set: 978-2-35158-207-7, Vol.1: 978-2-35158-209-1, Vol.2: 978-2-35158-210-7, e-ISBN: 978-2-35158-206-0; *Ed. Alexandra Bertron*

PRO 124 (online version): Fourth International Conference Progress of Recycling in the Built Environment (2018), e-ISBN: 978-2-35158-208-4; *Eds. Isabel M. Martins, Carina Ulsen, Yury Villagran*

PRO 125 (online version): SLD4—4th International Conference on Service Life Design for Infrastructures (2018), e-ISBN: 978-2-35158-213-8; *Eds. Guang Ye, Yong Yuan, Claudia Romero Rodriguez, Hongzhi Zhang, Branko Savija*

PRO 126: Workshop on Concrete Modelling and Material Behaviour in honor of Professor Klaas van Breugel (2018), ISBN: 978-2-35158-214-5, e-ISBN: 978-2-35158-215-2; *Ed. Guang Ye*

PRO 127 (online version): CONMOD2018—Symposium on Concrete Modelling (2018), e-ISBN: 978-2-35158-216-9; *Eds. Erik Schlangen, Geert de Schutter, Branko Savija, Hongzhi Zhang, Claudia Romero Rodriguez*

PRO 128: SMSS2019—International Conference on Sustainable Materials, Systems and Structures (2019), ISBN: 978-2-35158-217-6, e-ISBN: 978-2-35158-218-3

PRO 129: 2nd International Conference on UHPC Materials and Structures (UHPC2018-China), ISBN: 978-2-35158-219-0, e-ISBN: 978-2-35158-220-6

PRO 130: 5th Historic Mortars Conference (2019), ISBN: 978-2-35158-221-3, e-ISBN: 978-2-35158-222-0; *Eds. José Ignacio Álvarez, José María Fernández, Íñigo Navarro, Adrián Durán, Rafael Sirera*

PRO 131 (online version): 3rd International Conference on Bio-Based Building Materials (ICBBM2019), e-ISBN: 978-2-35158-229-9; *Eds. Mohammed Sonebi, Sofiane Amziane, Jonathan Page*

PRO 132: IRWRMC'18—International RILEM Workshop on Rheological Measurements of Cement-based Materials (2018), ISBN: 978-2-35158-230-5, e-ISBN: 978-2-35158-231-2; *Eds. Chafika Djelal, Yannick Vanhove*

PRO 133 (online version): CO2STO2019—International Workshop CO2 Storage in Concrete (2019), e-ISBN: 978-2-35158-232-9; *Eds. Assia Djerbi, Othman Omikrine-Metalssi, Teddy Fen-Chong*

PRO 134: 3rd ACF/HNU International Conference on UHPC Materials and Structures—UHPC'2020, ISBN: 978-2-35158-233-6, e-ISBN: 978-2-35158-234-3; *Eds. Caijun Shi and Jiaping Liu*

PRO 135: Fourth International Conference on Chemically Activated Materials (CAM2021), ISBN: 978-2-35158-235-0, e-ISBN: 978-2-35158-236-7; *Eds. Caijun Shi and Xiang Hu*

RILEM Reports (REP)

Report 19: Considerations for Use in Managing the Aging of Nuclear Power Plant Concrete Structures (ISBN: 2-912143-07-1); *Ed. D. J. Naus*

Report 20: Engineering and Transport Properties of the Interfacial Transition Zone in Cementitious Composites (ISBN: 2-912143-08-X); *Eds. M. G. Alexander, G. Arliguie, G. Ballivy, A. Bentur and J. Marchand*

Report 21: Durability of Building Sealants (ISBN: 2-912143-12-8); *Ed. A. T. Wolf*

Report 22: Sustainable Raw Materials—Construction and Demolition Waste (ISBN: 2-912143-17-9); *Eds. C. F. Hendriks and H. S. Pietersen*

Report 23: Self-Compacting Concrete state-of-the-art report (ISBN: 2-912143-23-3); *Eds. Å. Skarendahl and Ö. Petersson*

Report 24: Workability and Rheology of Fresh Concrete: Compendium of Tests (ISBN: 2-912143-32-2); *Eds. P. J. M. Bartos, M. Sonebi and A. K. Tamimi*

Report 25: Early Age Cracking in Cementitious Systems (ISBN: 2-912143-33-0); *Ed. A. Bentur*

Report 26: Towards Sustainable Roofing (Joint Committee CIB/RILEM) (CD 07) (e-ISBN 978-2-912143-65-5); *Eds. Thomas W. Hutchinson and Keith Roberts*

Report 27: Condition Assessment of Roofs (Joint Committee CIB/RILEM) (CD 08) (e-ISBN 978-2-912143-66-2); *Ed. CIB W 83/RILEM TC166-RMS*

Report 28: Final report of RILEM TC 167-COM 'Characterisation of Old Mortars with Respect to Their Repair (ISBN: 978-2-912143-56-3); *Eds. C. Groot, G. Ashall and J. Hughes*

Report 29: Pavement Performance Prediction and Evaluation (PPPE): Interlaboratory Tests (e-ISBN: 2-912143-68-3); *Eds. M. Partl and H. Piber*

Report 30: Final Report of RILEM TC 198-URM 'Use of Recycled Materials' (ISBN: 2-912143-82-9; e-ISBN: 2-912143-69-1); *Eds. Ch. F. Hendriks, G. M. T. Janssen and E. Vázquez*

Report 31: Final Report of RILEM TC 185-ATC 'Advanced testing of cement-based materials during setting and hardening' (ISBN: 2-912143-81-0; e-ISBN: 2-912143-70-5); *Eds. H. W. Reinhardt and C. U. Grosse*

Report 32: Probabilistic Assessment of Existing Structures. A JCSS publication (ISBN 2-912143-24-1); *Ed. D. Diamantidis*

Report 33: State-of-the-Art Report of RILEM Technical Committee TC 184-IFE 'Industrial Floors' (ISBN 2-35158-006-0); *Ed. P. Seidler*

Report 34: Report of RILEM Technical Committee TC 147-FMB 'Fracture mechanics applications to anchorage and bond' Tension of Reinforced Concrete Prisms—Round Robin Analysis and Tests on Bond (e-ISBN 2-912143-91-8); *Eds. L. Elfgren and K. Noghabai*

Report 35: Final Report of RILEM Technical Committee TC 188-CSC 'Casting of Self Compacting Concrete' (ISBN 2-35158-001-X; e-ISBN: 2-912143-98-5); *Eds. Å. Skarendahl and P. Billberg*

Report 36: State-of-the-Art Report of RILEM Technical Committee TC 201-TRC 'Textile Reinforced Concrete' (ISBN 2-912143-99-3); *Ed. W. Brameshuber*

Report 37: State-of-the-Art Report of RILEM Technical Committee TC 192-ECM 'Environment-conscious construction materials and systems' (ISBN: 978-2-35158-053-0); *Eds. N. Kashino, D. Van Gemert and K. Imamoto*

Report 38: State-of-the-Art Report of RILEM Technical Committee TC 205-DSC 'Durability of Self-Compacting Concrete' (ISBN: 978-2-35158-048-6); *Eds. G. De Schutter and K. Audenaert*

Report 39: Final Report of RILEM Technical Committee TC 187-SOC 'Experimental determination of the stress-crack opening curve for concrete in tension' (ISBN 978-2-35158-049-3); *Ed. J. Planas*

Report 40: State-of-the-Art Report of RILEM Technical Committee TC 189-NEC 'Non-Destructive Evaluation of the Penetrability and Thickness of the Concrete Cover' (ISBN 978-2-35158-054-7); *Eds. R. Torrent and L. Fernández Luco*

Report 41: State-of-the-Art Report of RILEM Technical Committee TC 196-ICC 'Internal Curing of Concrete' (ISBN 978-2-35158-009-7); *Eds. K. Kovler and O. M. Jensen*

Report 42: 'Acoustic Emission and Related Non-destructive Evaluation Techniques for Crack Detection and Damage Evaluation in Concrete'—Final Report of RILEM Technical Committee 212-ACD (e-ISBN: 978-2-35158-100-1); *Ed. M. Ohtsu*

Report 45: Repair Mortars for Historic Masonry—State-of-the-Art Report of RILEM Technical Committee TC 203-RHM (e-ISBN: 978-2-35158-163-6); *Eds. Paul Maurenbrecher and Caspar Groot*

Report 46: Surface delamination of concrete industrial floors and other durability related aspects guide—Report of RILEM Technical Committee TC 268-SIF (e-ISBN: 978-2-35158-201-5); *Ed. Valerie Pollet*

Contents

Chapter 1
Introduction

Lily D. Poulikakos, Emiliano Pasquini, Augusto Cannone Falchetto, and Di Wang

Abstract By promoting waste reuse and recycling and gradually banning waste disposal or incineration, a "circular economy" converts products that have reached the end of their service life into resources for the same or different industries, improving resource efficiency and reducing the effects on the climate and environment. Road pavements have been successfully constructed using a variety of waste materials, including plastic, crumb rubber, steel slag, and construction and demolition waste, which indicate positive environmental and economic effects. However, based on the data currently available, this is mainly at the research level or restricted to a few nations; as a result, solutions need to be developed and widely demonstrated to encourage the market uptake of such alternative materials. TC 279-WMR aimed to facilitate the use of waste, marginal and secondary materials for roads by investigating the performance of road materials containing waste through round-robin tests and the development of standard procedures for their selection, preparation, and implementation.

Keywords Asphalt · Waste · Polyethylene · Crumb rubber · Steel slag · Recycled concrete aggregates · LCA

L. D. Poulikakos (✉)
EMPA, Swiss Federal Laboratory for Material Science and Technology, Überlandstrasse 129, 8600 Dübendorf, Switzerland
e-mail: lily.poulikakos@empa.ch

E. Pasquini
Department of Civil, Environmental and Architectural Engineering, University of Padova, Via Marzolo 9, 35131 Padova, Italy

A. Cannone Falchetto · D. Wang
Department of Civil Engineering, Aalto University, Rakentajanaukio 4, 02150 Espoo, Finland

© The Author(s), under exclusive license to Springer Nature Switzerland AG 2023
A. Cannone Falchetto et al. (eds.), *Valorisation of Waste and Secondary Materials for Roads*, RILEM State-of-the-Art Reports 38,
https://doi.org/10.1007/978-3-031-33173-2_1

1 Background and Motivation

Turning products that have reached the end of their service lives into resources for the same or new industries is the ideal application of a "circular economy." By encouraging waste reuse and recycling and gradually eliminating waste disposal or incineration, this approach has the potential to increase resource efficiency and lessen climatic and environmental impacts. However, the linear process of material usage in our economies, in which raw materials are extracted, consumed, or further processed into products where they are used, and then disposed of in landfills and incinerators, means that a large portion of this resource is wasted nowadays. Most waste materials were recycled into the same kinds of things that they originally were (closed-loop recycling). Nevertheless, despite significant advancements in the recycling of waste materials, in many instances, their technical or quality specifications prevent the resources from being recycled into the same product. The leftover materials are consequently burned or dumped. An important opportunity could lie in the recycling of some waste materials as precious secondary raw materials for different applications, such as for example in road pavements. As asphalt surface courses account for more than 90% of the paved surfaces in Europe [1], resulting in the creation of large amounts of asphalt mixtures and the substantial use of nonrenewable natural aggregates and asphalt binder. However, currently, the main recycled material used in roads is Reclaimed Asphalt Pavement (RAP) [2]. Current international research has shown that various types of waste materials such as plastic, crumb rubber, steel slag, and construction and demolition waste have been successfully used in road pavements leading to environmental and economic benefits [3]. However, mostly at the academic level or restricted to certain nations/regions. To encourage extensive market uptake, it is urgently necessary to develop and broadly showcase such solutions. Piao et al. [4] have looked at several waste materials used for roads. The current experience indicates that when considering rutting, moisture, stiffness, and fatigue performance, these waste materials had similar or better performance to reference mixtures made of all virgin materials. At the same time, the technology readiness level (TRL) varied considerably worldwide. Crumb rubber from tires is one representative example in this sense since its use is quite widespread in some parts of the world, while others only have experience with this waste material in the laboratory. The reasons for this difference can be listed as follows: (i) lack of knowledge for their use, (ii) lack of trust in their performance, (iii) lack of standards, (iv) lack of incentives.

RILEM technical committees offer one of the best platforms for international cooperation to handle all aspects pertaining to the utilization of waste and marginal materials for roads. This can include everything from material selection to laboratory characterization to in-situ performance of chosen materials. This technical committee established a solid and essential framework for the assessment of the utilization of diverse waste and marginal products as performance-reinforcing substitutes for conventional road materials.

TC 279-WMR was closely linked to TC 231-NBM (Nano-Technology Based Bituminous Materials), TC 237-SIB (Testing and Characterization of Sustainable Innovative Bituminous Materials and Systems), TC 241-MCD (Mechanics of Cracking in Asphalt and Composite Pavements), TC 252 CMB (Chemo-Mechanical Characterization of Bituminous Materials), TC 264-RAP (Recycling of Asphalt Pavements), and TC 272-PIM (Phase and Interphase behavior of innovative bituminous materials). As well as existing national and international associations (e.g., AAPA, AAPT, AASHTO, ACI, ARRA, ASTM, CEN, EAPA, Eurobitume, FEHRL, ISAP, ISO, NAPA, PIARC, TRB). In this respect, TC 279-WMR continued to profit from the existing relationships and contributed to this already well-established international network.

This TC fits perfectly into the RILEM's technical program, particularly the items "Materials Characterization; Properties Evaluation and Processing; Performance and Deterioration Mechanisms; Special Construction Materials and Components; and Mechanical Performance and Fracture." The TC 279-WMR had its annual meeting at the same place and in series with TC 241-MCD and TC 264-RAP, TC 272-PIM to ensure complementary work, avoid repetition and encourage cross-fertilization of information. The goals of this TC could be realized, and solutions could be sought only through interdisciplinary research.

This TC aimed to add to the state of the art regarding the use of waste, marginal and secondary materials for roads by investigating their performance in road materials containing waste through round-robin tests and the development of standard procedures for their selection, preparation, and use.

The following items were the primary aims of TC 279-WMR:

(i) Identify waste, marginal, and secondary materials that can be used in road materials to improve performance and/or durability
(ii) Evaluate the effectiveness of the modified binders and characterize the performance of appropriate binder additions
(iii) Study the performance of possible alternatives for aggregates in mixtures
(iv) Propose appropriate wastes materials that can be used in road and set the maximum content/amount on their usage
(v) Assess the sustainability of using waste materials by appropriate life cycle analysis.

2 Structure of the TC

Due to its multidisciplinary nature, TC 279-WMR attracted a wide pallet of members such as civil engineers, material scientists, waste material producers, road authorities, chemists, test laboratories, and test equipment producers. This TC incorporated also members from outside Europe, such as Canada, India, US, and Brazil, to broaden the global platform. The results of this TC can be equally interesting to road authorities and waste management professionals.

The TC was active for five years (2017–2022). A combination of literature review and experimental work was done within four task groups.

2.1 TG1 and TG2: Asphalt Binder Additives

Waste products of various kinds can be used as asphalt binders' performance-enhancing additives. TG1 evaluated the impact of polyethylene on asphalt binders, while TG2 assessed the effect of end-of-life tire-produced crumb rubbers. Various methods described below were used to assess the use of binder additives and their effects on the material's physical, chemical, and mechanical qualities.

Physical–chemical characterization was performed using Fourier transform infrared spectroscopy (FTIR) and Differential Scanning Calorimetry (DSC). FTIR was used to determine changes in the functional groups of the modified binders. To demonstrate any changes in the glass transition temperature and the melting temperature that might have an impact on the mechanical behavior, DSC was used. Additionally, environmental scanning electron microscopy (ESEM) was used to examine the waste materials' microstructure.

For the purpose of identifying potential key behaviors, the characteristics of the various virgin and waste-modified asphalt binders were determined by using both conventional rheological characterization methods and more sophisticated approaches. In order to accomplish this target, a Dynamic Shear Rheometer (DSR) was employed to evaluate the materials' response over a wide temperature range. Advanced multi-scaling techniques were also used in experiments to assess the low and high temperatures, and fatigue behavior of asphalt binders containing waste materials. In order to validate the final waste-modified products, limitations related to extraction and recovery techniques were also discussed, and solutions for routine forensic and quality control testing in asphalt research institutions, road authorities, and the industrial partner were presented.

2.2 TG3: Aggregate and Filler Substitutes

The following materials were investigated in TG3 as potential replacements for conventional nonrenewable aggregates, including Construction and Demolition Waste (CDW), Recycled Concrete Aggregates (RCA), and Steel Slag (SS).

Testing and ranking the waste aggregates: selected wastes were experimentally measured to evaluate their mechanical, chemical, and mineralogical properties before being ranked in accordance with the standards currently used for granular pavement materials. These characteristics included bulk density, shape, resistance to fragmentation, strength, water absorption, volume stability, and abrasion resistance. The suitability of the substitute materials could be evaluated within this subtask.

In this subtask of mixtures' mechanical performance, different proportions of each recycled waste material were explored and compared to the behavior of asphalt mixtures prepared with all virgin materials using mechanical performance characterization. The testing plan included water sensitivity, stiffness, fatigue resistance, permanent deformation resistance, and low temperature behavior of asphalt mixtures. In cooperation with TG1, this TG also assessed the feasibility of using waste polyethylene in asphalt mixtures as dry addition to the virgin aggregates. Such supplementary work was carried out following the same approach described above.

2.3 *TG4: Life Cycle Assessment*

TG4 provided pavement specialists with international average values of greenhouse gas emissions (GHG) and energy consumption related to the manufacturing of asphalt mixtures, with and without waste and secondary materials for roads (WMRs). In order to achieve this goal, current standards to define a methodology for obtaining benchmark values were used. Thereafter, a data collection tool was designed and provided to a group of international partners to build a life cycle inventory for asphalt mixtures including data from material supply up to plant manufacturing. This set of results was used to obtain benchmark values for reference (conventional) asphalt mixtures, namely Asphalt Concretes (ACs) and Stone Matrix Asphalts (SMAs). Finally, in order to assess how environmentally-friendly it is to incorporate WMRs into asphalt mixtures, the benchmark values of the reference asphalt concrete and stone matrix asphalt mixtures were compared using the GHG and energy consumptions of asphalt mixtures manufactured with selected WMRs.

3 Scope of This Book

The work of TC 279-WMR has supplied the scientific foundation for utilizing a portion of the waste and marginal materials produced annually, which would otherwise be burned or disposed of, to be used in asphalt concrete. Thereby resulting in potentially considerable favorable effects on the environment, the economy, and society. The findings can be utilized to increase awareness on this topic and as a foundation to enhance national and international standards to make it easier to use such resources and moving towards a zero waste society. To this end, such findings can be utilized to create a global framework for additional advancement in this area. The State-of-the-Art Report, recommendations, and publications that were created will also serve as a solid foundation for the education of upcoming scientists and engineers, making them a significant investment for the future.

References

1. EAPA, Learn about asphalt. https://eapa.org/asphalt/. Accessed 11 May 2022
2. EAPA, Asphalt paving—fit for tomorrow (2011): safe, comfortable, sustainable. https://eapa.org/asphalt-paving-fit-for-tomorrow-2011/. Accessed 11 May 2022
3. Poulikakos LD, Papadaskalopoulou C, Hofko B, Gschösser F, Cannone Falchetto A, Bueno M et al (2017) Harvesting the unexplored potential of European waste materials for road construction. Resour Conserv Recycl 116:32–44. https://doi.org/10.1016/j.resconrec.2016.09.008
4. Piao Z, Mikhailenko P, Kakar MR, Bueno M, Hellweg S, Poulikakos LD (2021) Urban mining for asphalt pavements: a review. J Clean Prod 280:124916. https://doi.org/10.1016/j.jclepro.2020.124916

Chapter 2
Bituminous Binder and Bituminous Mixture Modified with Waste Polyethylene

Marjan Tušar, Lily D. Poulikakos, Muhammad Rafiq Kakar, Emiliano Pasquini, Marco Pasetto, Laurent Porot, Di Wang, Augusto Cannone Falchetto, Alan Carter, Gabriel Orozco, Chiara Riccardi, Kamilla Vasconcelos, Aikaterini Varveri, Ruxin Jing, Gustavo Pinheiro, David Hernando, Peter Mikhailenko, Jan Stoop, Lacy Wouters, Miomir Miljković, Marko Orešković, Nunzio Viscione, Rosa Veropalumbo, Nikhil Saboo, Éric Lachance-Tremblay, Michel Vaillancourt, Nicolas Bueche, Davide Dalmazzo, Fernando Moreno-Navarro, Davide Lo Presti, and Gaspare Giancontieri

Abstract RILEM TC-279 WMR task group TG 1 studied the performance of waste Polyethylene (PE) in bituminous binders and bituminous mixtures. Several laboratories participated in this study following a common protocol. Locally sources aggregates and bituminous binder and same source of waste PE were utilized. The

M. Tušar (✉)
Slovenian National Building and Civil Engineering Institute, Dimičeva ulica 12, 1000 Ljubljana, Slovenia
e-mail: marjan.tusar@ki.si

National Institute of Chemistry, Hajdrihova 19, 1001 Ljubljana, Slovenia

L. D. Poulikakos · M. R. Kakar · P. Mikhailenko
Empa, Swiss Federal Laboratories for Materials Science and Technology, Überlandstrasse 129, 8600 Dübendorf, Switzerland

M. R. Kakar · N. Bueche
Department of Architecture, Wood and Civil Engineering, Bern University of Applied Sciences (BFH), Bern, Switzerland

E. Pasquini · M. Pasetto
Department of Civil, Environmental and Architectural Engineering (ICEA), University of Padova, Via Marzolo, 9, 35131 Padova, Italy

L. Porot
Kraton Polymer B.V., Transistorstraat 16, 1322 CE Almere, The Netherlands

D. Wang · A. Cannone Falchetto
Department of Civil Engineering, Aalto University, Rakentajanaukio 4, 02150 Espoo, Finland

© The Author(s), under exclusive license to Springer Nature Switzerland AG 2023

A. Cannone Falchetto et al. (eds.), *Valorisation of Waste and Secondary Materials for Roads*, RILEM State-of-the-Art Reports 38,
https://doi.org/10.1007/978-3-031-33173-2_2

binder experiments showed that at high temperatures, using MSCR tests, PE modified blends had better resistance to permanent deformation in comparison to the non modified binder. Whereas at intermediate temperatures, using the LAS tests, fatigue performance of the PE blends could withstand more loading cycles under low strains; however, it could sustain less loading cycles under high strains due to the increase in brittleness. Dry process was used for the mixture experiments in order to bypass the stability and inhomogeneity experience that was observed at the binder scale. The PE modified mixtures showed improved workability and increased strength. The higher the PE dosage, the higher the ITS increase with respect to the values measured for

A. Carter · G. Orozco · É. Lachance-Tremblay · M. Vaillancourt
Department of Construction Engineering, École de Technologie Supérieure (LCMB-ETS), Montreal, QC H3C, Canada

C. Riccardi
Department of Civil and Industrial Engineering, University of Pisa, Largo L. Lazzarino, 1, 56122 Pisa, Italy

K. Vasconcelos · G. Pinheiro
Escola Politécnica da Universidade de São Paulo, Av. Prof. Luciano Gualberto, 380, Butantã, São Paulo - SP 05508-010, Brazil

A. Varveri · R. Jing
Section of Pavement Engineering, Faculty of Civil Engineering and Geosciences, Delft University of Technology, Stevinweg 1, 2628 CN Delft, The Netherlands

D. Hernando · J. Stoop · L. Wouters
Energy and Materials in Infrastructure and Buildings (EMIB), Faculty of Applied Engineering, University of Antwerp, Antwerp, Belgium

M. Miljković
Faculty of Civil Engineering and Architecture, University of Niš, Aleksandra Medvedeva 14, 18000 Niš, Serbia

M. Orešković
Faculty of Civil Engineering, University of Belgrade, Bulevar kralja Aleksandra 73, 11000 Belgrade, Serbia

N. Viscione · R. Veropalumbo
Department of Civil, Construction and Environmental Engineering, University of Naples Federico II, Naples, Italy

N. Saboo
Indian Institute of Technology Roorkee, Varanasi, India

D. Dalmazzo
Department of Environment, Land and Infrastructure Engineering, Politecnico di Torino, Torino, Italy

F. Moreno-Navarro
LABIC-UGR, Laboratorio de Ingeniería de la Construcción Universidad de Granada ETS Caminos, Canales y Puertos Severo Ochoa S/N Campus de Fuentenueva, 18071 Granada, Spain

D. Lo Presti · G. Giancontieri
University of Palermo, Piazza Marina, 61, 90133 Palermo, Italy

the control materials (i.e., without any plastic waste) thanks to the improved cohesion of the plastic modified mastic. The stiffness experiments tended to show an improved performance with a lower time dependence and a higher elasticity when plastic was added. The cyclic compression tests demonstrated a reduced creep rate along with a higher creep modulus thanks to the addition of PE; similar conclusions can be drawn from the experimental findings coming from wheel tracking test. Furthermore, acceptable and often improved moisture resistance was observed for PE modified materials.

Keywords Binder · Asphalt · Bituminous mixture · Waste · Polyethylene

1 Introduction

Discovery of plastic materials in the beginning of the twentieth century enabled acceleration in the development of the human civilization. In 1839, Charles Goodyear discovered the process of making plastics by chemically modifying natural polymers—the process of vulcanizing rubber—forming connections between polymer chains of rubber. In 1909, Leo Hendrik Baekeland created Bakelite. Fully synthetic plastic material made of phenol and formaldehyde. By modifying this process later polyvinyl chloride, polystyrene, polyethylenes, polypropylene, polyamide (nylon), polyester, acrylic, silicone and polyurethane were produced. In 1910, Sergei Lebedev synthesized artificial rubber, which led to the development of styrene-butadiene rubber during World War II. Since then, plastic materials are widely used all over the world. However, as plastic materials are cheap and widely used, in the last 50 years enormous production of cheap plastic materials has unfortunately led to large amounts of plastic waste.

Following the principles of a sustainable development, and a circular economy goods need to be recycled or reused at the end of their service life [1]. International research results have shown that many waste materials, including waste plastics, can be successfully used as alternative materials to the virgin materials in roads [2–4]. It was documented by Piao et al. [3], that the Technology Readiness Level (TRL) for the use of waste materials for road construction for many of the viable materials has remained, for the most part, at the research level or limited to use of some materials and countries. This indicates that in order to open up a broad market acceptance, such solutions must be developed and widely demonstrated. The main goal of the present work was to evaluate the use of such alternatives materials for use in pavements and as a result to develop a robust and fundamental understanding for their use, with the aim of delivering at least similar performances to conventional materials. To achieve this target, inter-laboratory tests were used to evaluate the performance of road materials that contained wastes, and standard processes for their selection, preparation, and usage were developed. Several types of waste plastics were considered at the beginning of the work of TG 1. However, it was decided to perform experimental work only with waste Polyethylene (PE) from one source, due to the fact that the softening point of PE is around 125 °C which is lower than the temperature of production of hot bituminous mixtures (around 150 °C). It was expected that PE would be easily and evenly mixed in the bituminous binder and in the bituminous mixture. In this work, two types of PE applications were performed: first, waste PE was mixed with bituminous binder and with waste PE mixed in bituminous mixture according to dry procedure.

2 Literature Review

Various test methods were selected for the characterization of PE modified binders. The literature review below is focused on these tests namely Temperature frequency sweep (*T-f*-sweep), Multiple Stress Creep and Recovery (MSCR), Linear Amplitude Sweep (LAS) rely on Dynamic Shear Rheometer (DSR), Differential scanning calorimetry (DSC), and Fourier Transform Infrared Spectroscopy (FTIR). To characterize the PE modified mixtures workability, strength, stiffness and permanent deformation were investigated. The background of these testing methods will be explained below.

The Dynamic Shear Rheometer (DSR) [5] is frequently used to assess the viscoelastic characteristics of bituminous binders at high and intermediate temperatures. Recent studies also have indicated that the 4 mm DSR can also be utilized for low temperature characterization [6, 7]. In general, a combination of multiple experimental configurations, such as sample geometries, temperatures, frequencies, complicated strain/stress levels, and static/dynamic loading mode, were applied to assess the rheological properties of bituminous binders by using DSR [5, 8]. The *T-f-sweep* test findings have been proven to be one of the most effective instruments for evaluating the rheological characteristics of bituminous binders and differentiating between non modified and modified binders over a wide range of temperatures and frequencies. The measured data, such as $|G^*|$ and δ, can be further evaluated in accordance with the Time–Temperature Superposition Principle (TTSP) [9, 10], such as by creating master curves and figuring out the associated rheological parameters. These include the crossover temperature [11, 12], the Glover-Rowe parameter [13, 14], the rheological index R [15], and $\triangle T_c$ [16, 17].

Nowadays, researchers employed DSR to assess how PE additions affected modified binders. Compared to the reference non modified binders, the binder modified with waste plastic shows generally improved rheological characteristics [18]. Enhanced viscoelastic characteristics, including higher $|G^*|$ and reduced δ [19–22], were discovered. However, high temperatures also revealed poor repeatability and reproducibility [23].

The Multiple Stress Creep and Recovery (MSCR) test, included in the Superpave Performance Graded (PG) specification, aims to characterize the rutting resistance of bituminous binders under application of different stress levels, and enables to assess both the non-recoverable creep compliance (J_{nr}) and the percent recovery (R) parameters [24]. J_{nr} can be calculated by the residual strain in a sample after one creep and recovery cycle divided by the stress applied (kPa^{-1}), while R (%) represents the elastic response and stress dependence of the bituminous binder. The plot of R related to J_{nr} can also identify the presence of a polymer and its crosslinking density as an alternative method for other tests [25].

Recently, several studies used the MSCR tests to evaluate the anti-rutting properties of binders modified with waste PE [26–30]. These investigations reported benefits regarding the elastic response with the addition of PE up to 6% by weight of bituminous binder, lowering the composite's unrecovered strain under high temperatures. At 58 °C and stress level of 3.2 kPa, for example, PE addition provided results of J_{nr} between 0.24 and 0.51 kPa^{-1}, while non-modified bituminous binder presented values between 1.02 and 1.75 kPa^{-1} [26, 29]. This favorable condition is justified because of the recycled thermoplastics high ductility and strength properties, even when submitted to high service temperatures.

Additionally, the stress sensitivity of the binders modified with waste PE was analyzed based on the percent difference in non-recoverable compliances ($J_{nr,diff}$). Zhou et al. [30] indicated that reduced stress sensitivity with the presence of PE in the bituminous binder. However, this trend was not observed by Nuñez et al. [29]. Nevertheless, all of the tested binders demonstrated reliable stress sensitivity in relation to the specification's maximum limiting value of 75% [31].

Joohari and Giustozzi [27] investigated specifically the effects of adding different types of PE to an SBS (styrene–butadiene–styrene) modified bituminous binder, e.g., low-density Polyethylene (LDPE), linear low-density Polyethylene (LLDPE), and a recycled linear low-density Polyethylene from waste plastics (RLLDPE). The study concluded that the addition of these thermoplastics can increase rutting susceptibility, based on the MSCR test parameters. More studies should be carefully performed in order to assess the impact of additional plastic types, with or without cross-linking agents' incorporation (e.g., sulfur) aiming to enhance the polymers stabilization.

Developed to assess the bituminous binders fatigue performance, the Linear Amplitude Sweep (LAS) test applies cyclic loads with increasing amplitudes to accelerate damage in order to assess the ability of bituminous binders to endure fatigue damage [32]. Usually, this behavior is interpreted based on two analyses, the viscoelastic continuum damage (VECD) model with test results of both parameters A_{35} and B, and the damage tolerance parameter, α_f. The material characteristics against accumulated damage are represented by the parameter A_{35}, while the material sensitivity to changes of the applied shear loads is represented by the parameter B.

Researches, that evaluated fatigue resistance of bituminous binder modified with PE, reported benefits when PE is incorporated into neat bituminous binder, indicating higher values of fatigue life (N_f) independent of the shear strain levels tested and the type of PE [26–28]. Overall, tested bituminous binder samples were submitted to short- and long-term aging, with the exception of Zhou et al. [30] and Joohari and Giustozzi [27], that used unaged bituminous binders. Bituminous binders modified with PE are mostly less sensitive and more resistant to fatigue cracking [30].

Differently, Nuñez et al. [29] observed that, for higher shear strain levels (higher than 7%), the binder modified with PE presented worse fatigue performance than the reference neat bituminous binder, for all tested temperature and aging conditions, respectively, at 25 and 35 °C, and both short ASTM D2872-04 [31], EN 12607-1 [32] and long-term aging, ASTM D6521-08 [33], EN 14769 [34]. Thus, further investigations regarding the fatigue behavior at high strain levels should be evaluated, since only two of these studies considered strain levels beyond 10%.

A thermo analytical technique named Differential Scanning Calorimetry (DSC) enables the identification of physical changes brought about by heating and cooling in a substance. As DSC may be used to measure a variety of characteristics and phenomena in the material, therefore, it has been employed extensively in the analysis of bituminous binder [35]. Identification of endothermic or exothermic activities is aided by the DSC analysis. Measurements are made of important characteristics including the glass transition temperature (T_g), and phase transitions like melting and crystallization. Additionally, it is possible to measure chemical processes and heat capacity [36, 37]. The findings might be used to determine the wax content, demonstrate the presence of wax in bituminous binders, and the melting point of the wax [38]. Bituminous material containing many different species, the determination of T_g is not straightforward and may require additional measurement features including modulated transition [39]. Using DSC technique in modified bituminous binder is a powerful tool to identify the morphological structure of the binder. In a perfect blend the glass transition will result in an intermediate T_g, while in a two-phase binder, the glass transitions of both compounds remain distinct.

In order to characterize chemical changes in the binders as a result of the blending with PE, Attenuated Total Reflectance Fourier Transform Infrared Spectroscopy (ATR-FTIR) was used. The ATR-FTIR method for bituminous binder allows to identify certain functional groups. With the FTIR technique, the molecules are excited by electromagnetic waves in the infrared range of 4000–400 cm^{-1}. Molecular functional groups will absorb specific electromagnetic radiation leading to vibration and increase in total energy. With bituminous binder the most interesting wavelengths are located between 1800 and 600 cm^{-1} to track structure and modifications. It is often used in research to characterize bituminous binders [40, 41]. Especially to address modifications either occurring during aging through oxidation [42] or from blending with specific additives or modifiers [43]. With proper analysis of the spectra it is possible to define typical footprint of conventional bituminous binder [44], which later can help in identifying different modifiers through specify peaks in the spectrum.

Bituminous mixtures can be characterized by various laboratory methods as a means to predict their performance during their service life. The particular methods should be able to identify weaknesses in the material. For example, workability of the material can be assessed by analyzing the gyratory compaction curves [45]. The tensile strength [46] of the material can be determined in both dry and wet conditions and used to evaluate the strength and the moisture resistance, respectively. On the other hand, stiffness and linear visco-elastic characteristics can be studied through pulse indirect tension tests and cyclic tension–compression tests at intermediate service temperatures (25–35 °C). Finally, Cyclic Compression Tests (CCT) [47] on cylindrical specimens or Wheel Tracking Tests (WTT) [48] on slabs can be used to analyze the resistance of the investigated materials against permanent deformation at 60 °C.

Table 1 Detailed compositions of the investigated binders

Laboratories		1	2	3	4	5	6	7	8	9	10	11
B			√					√			√	√
Batch 1 (2018)	$B_{+pellets}$		√									
	$B_{+shreds}$	√	√	√	√	√	√			√		√
Batch 2 (2019)	$B_{+pellets}$	√						√	√		√	
	$B_{+shreds}$	√						√	√		√	

3 Bituminous Binder Modified with Waste PE

3.1 Preparation of Modified Bituminous Binder

The materials used for this inter-laboratory study were straight-run bituminous binder 70/100 with a penetration of 82 mm^{-1} and softening point of 50 °C and two types of polyethylene waste (pellets and shreds). The pellets were produced from waste packaging and the shreds were a secondary waste as a result of the pellets production. The waste PE pellets and shreds were ground to smaller sizes of ca. 2 mm before being blended with bituminous binder. A single 5% content of PE additive was used to prepare the blended binders, for the blending procedure the PE waste was added to hot bituminous binder that was heated to 170 °C for 1 h and mixed with a speed of 3500 rpm using a high shear mixer for 1 h. More details on the process of grinding and blending can be found at Kakar et al. [49]. Three types of blends were used in this work with the following designations: *B* (bituminous binder B 70/100); $B_{+pellets}$ (bituminous binder B 70/100 blended with PE-pellets), and $B_{+shreds}$ (bituminous binder B 70/100 blended with PE-shreds). The detailed composition of the investigated binders was displayed in Table 1.

3.2 Experimental Plan

Inter laboratory tests were conducted by laboratories located all over the world. The penetration values and softening point temperature were conducted based on the EN 1426 [50] and EN 1427 [51], respectively. The storage stability was performed based on the EN 13399 [52]. The rheological characterization of the bituminous binder was conducted with the Dynamic Shear Rheometer (DSR) [5]. Temperature-frequency sweep (*T-f*-sweep) tests were used to assess the materials' viscoelastic behaviour over a wide range of temperatures and frequencies. More specifically, for complex binders, the DSR measurements return engineering parameters such as complex modulus, G^*, that provide more insight into the material response than conventional physical properties used for purchase specification. In this study, the *T-f*-sweep tests were run by eleven laboratories with two standard geometries: 8 mm diameter and

2 mm gap for intermediate temperatures ranging from −6 to 40 °C (−6, 0, 4, 10, 16, 22, 28, 33, and 40 °C), and 25 mm diameter 1 mm gap for high temperatures from 33 to 82 °C with an interval of 6 °C. The imposed frequencies ranged from 0.1 to 20 Hz. Among the eleven participating institutions, four laboratories tested bituminous binder B, five tested $B_{+pellets}$, while the entire interlaboratory team worked on $B_{+shreds}$. More information can be accessed in the authors' previous publications [23, 53, 54].

The MSCR test [55, 56] is intended to assess the bituminous binder's sensitivity on permanent deformation. For this test, shear stress is applied using an oscillating load for a period of one second followed by an unloading (recovery) phase of nine seconds. The test starts with the application of a low stress 0.1 kPa for 10 creep/recovery cycles, follows with an increased stress of 3.2 kPa and repeated for an additional 10 cycles.

In this study, the tests were carried out using the 25 mm parallel plates configuration with 1 mm gap at 60 °C. The tests were performed on the neat binder (B), the blend with the PE shreds ($B_{+shreds}$) and the blend with PE pellets ($B_{+pellets}$). The tests were performed on freshly prepared samples in contrast to the standard specification request (after Rolling Thin Film Oven, RTFO). At least two replicates were tested per binder type. After 10 cycles using Eq. 1, at each stress level the average percent recovery (R) is determined. Furthermore, the non-recoverable creep compliance (J_{nr}) is calculated by dividing the non-recoverable shear strain by the shear stress. The average of the J_{nr} values after 10 loading cycles at each stress level is calculated using Eq. 2.

$$R_\tau = \frac{1}{10}\sum\nolimits_{N=1}^{10} \frac{(\varepsilon_1^N - \varepsilon_{10}^N)}{\varepsilon_1^N} \times 100\% \quad (\%) \tag{1}$$

$$J_{nr}^N = \frac{\varepsilon_{10}^N}{\tau} (\text{kPa}^{-1}) \tag{2}$$

where,

τ the applied stress, 0.1 and 3.2 kPa;
ε_1^N the strain value at the end of creep portion (after 1 s) of each cycle;
ε_{10}^N the strain value at the end of the recovery phase (after 10 s) of *N-th* cycle.

The LAS test [57] is used to assess the materials' resistance to fatigue damage by applying cyclic loading at amplitudes that increase until the damage occurs. Data were collected each second as the shear strain was linearly varied from 0% to 30%. The parallel-plate setup with an 8 mm plate diameter and 2 mm gap was used to record the binder samples at 10 Hz and 20 °C. At least two replicates were tested per binder type. The samples did not undergo any short- or long-term ageing before testing. The material's accumulated damage rate served as a proxy for the bituminous binder blends' fatigue behavior.

The information on the undamaged material characteristics (represented by the parameters α and β; see Eq. 3) was determined based on the outcomes of frequency sweep tests [58] in order to complete the fatigue analysis. The following equation was used to apply a straight line best-fit to the data, with loading frequency on the horizontal axis and storage modulus on the vertical axis, in order to get the α and β parameters [59].

$$\log G'(f) = \frac{1}{\beta}\log(f) + \beta \tag{3}$$

where,

G' the storage modulus;
f the reduced frequency;
α, β fitting constants.

Damage intensity (D_t) in the bituminous binder sample was calculated using the following summation.

$$D_t \cong \sum_{i=1}^{t}\left[\pi\gamma^2\frac{(G''_{i-1} - G''_i)}{G''_{initial}}\right]^{\frac{\alpha}{1+\alpha}}(t_i - t_{i-1})^{\frac{1}{1+\alpha}} \tag{4}$$

where,

γ the shear strain;
G'' the loss modulus;
t the time.

Finally, the performance indicator for binder fatigue (N_f) can be expressed according to the following equation.

$$N_f = A(\gamma)^B \tag{5}$$

where,

γ the shear strain;
A, B the viscoelastic continuum damage (VECD) model coefficients that depend on the material properties.

To be specific, A is defined using damage accumulation corresponding to a 35% decrease from the initial loss shear modulus, and B is determined by the linear viscoelastic properties of binder. In general, more fatigue resistant binders tend to have higher A values and lower absolute B values.

FTIR was performed by two laboratories lab7 and lab8, but only one run the test on the base bituminous binder 70/100 and the two blends with plastic shreds and pellets. It was conducted in Attenuated Total Reflectance (ATR) between 600 and 4000 cm^{-1} with 64 scans and a resolution of 4 cm^{-1}. In this present study, FTIR spectroscopy was used to track how plastic pellets and plastic shreds affected the reference binder's chemical composition. The examination of B, $B_{+pellets}$ and $B_{+shreds}$ was conducted by using the Bruker Vector 22/Digilab BioRad FTS 6000 FTIR spectrometer.

DSC was performed by one lab Lab7 on linear amplitude heat flow. First the sample was heated up at 165 °C and stabilized for 5 min to remove any heating history. The sample was then chilled at a rate of 2 °C/min until it reached −60 °C, kept for 5 min and then reheated at the same rate to 165 °C. The thermal properties of B, $B_{+pellets}$ and $B_{+shreds}$ were analyzed by using DSC. The followed testing protocol was followed: the samples were first cooled from 25 to 20 °C and held at this temperature for 5 min. Then, they were heated to 200 °C and held for 5 min before being cooled once more to 20 °C and held at that temperature for 5 min before being heated to 200 °C. A single cooling and heating rate of 20 °C/min was applied for such processes. For the sake of reproducibility, each sample was tested at least three times.

4 Results and Analysis on Bituminous Binder Tests

4.1 Conventional Tests Results

The results of penetration values seem similar among different laboratories, a reduction of 50% was observed when the PE additives were blended. However, this is not true for the softening point temperature, consistent results were obtained among non modified binders, while the results measured in the binders modified with PE were quite different. In $\mathrm{B}_{+\mathrm{shreds}}$, the softening point temperature results ranged from 45 to 110 °C which led to a maximum difference of more than 60 °C [53]. This may be attributed to the inhomogeneous distribution of plastic particles in the modified binders [53, 54]. More detailed information and discussion about the storage stability can be accessed in the authors' previous publication [49, 53].

4.2 Interlaboratory Comparison of DSR T-F-Sweep Tests Bituminous Binder

The results of the DSR measurements conducted on the binder modified with PE are analyzed and discussed in this section. To assess the repeatability within labs and reproducibility across laboratories, the isochronal curves of the complex shear modulus, $|G^*|$, and the phase angle, δ, at the reference frequency of 1.59 Hz (10 rad/s), were first constructed. All three binder types indicate quite a high repeatability in the results, while the binders modified with PE had low reproducibility. This shows that the mixes are not as homogeneous as bituminous binders made for ordinary paving.

Figure 1 makes it clear that the testing temperature had a considerable and direct impact on the outcomes. Reduced variability was observed for all the investigated materials at relatively low testing temperatures (PP08). This temperature effect was less pronounced for the binders modified with PE ($B_{+pellets}$ and $B_{+shreds}$). Within this low-temperature range, these blends exhibited a complex shear modulus that was comparable to that of the reference unmodified binder. In addition, all laboratories were able to detect an increase in the complex shear modulus relative to the neat binder in the high temperature regime. A more elastic phase angle behavior was seen in the low temperature domain. The inclusion of plastics, on the other hand, caused the phase angle to decrease in the high temperature regime as compared to the neat binder, indicating a more elastic response. Interestingly, the differences between the findings from the various laboratories were larger at the high temperature range. Due to the weaker binder and lower viscosity at the higher testing temperatures, these findings may indicate that the plastic particles are really being assessed instead of the entire mix, exposing the sample's heterogeneity and resulting variances in results (depending on the amount of plastic particles in the binder samples). A larger diameter and smaller gap of the samples utilized for the high temperature study (PP25) might be another element enhancing such an impact. Regardless of the testing temperature, reasonable sample stability was seen when binder B, $B_{+pellets}$, and $B_{+shreds}$ mixes were tested.

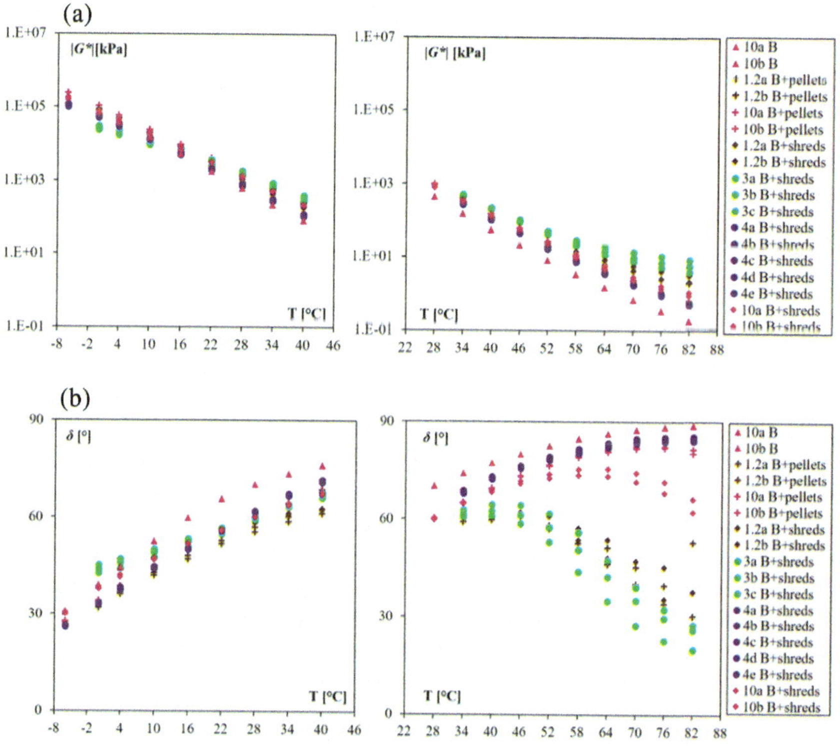

Fig. 1 Isochronal plots of binders modified with PE $B_{+pellets}$ and $B_{+shreds}$ at the reference frequency of 1.59 Hz (10 rad/s): **a** $|G^*|$ at low (left) and high (right) testing temperature; **b** δ at low (left) and high (right) testing temperature [53]

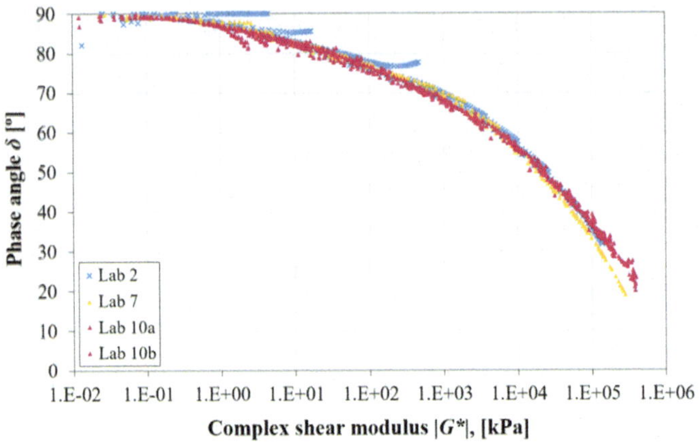

Fig. 2 Black diagram showing phase angle versus complex shear modulus for the neat bituminous binder 70/100 [53]

The analysis of the raw DSR data for all frequencies at the different temperatures was made using the black space with complex shear modulus, $|G^*|$, and phase angle, δ. This enables to display all data points for all frequencies and temperatures without any shift of the values.

Figure 2 shows the plot for the neat binder 70/100 as reported by the three laboratories in frequency sweep at different temperatures including two of them having performed also temperature ramping. Lab2 showed some discontinuity with the 25 mm geometry presuming morphology disruption during the test with no time temperature superposition. The other two labs gave reasonably good reproducibility either in frequency sweep or temperature ramping.

For shreds blends, eleven laboratories performed frequency sweep measurement. The analysis was divided in two parts first the 8 mm geometry representing the cold temperature regime and second the 25 mm geometry representing the warm temperature regime.

With the pellet blends the same observation was made with poor reproducibility as shown in Fig. 3. Figure 4a reports the measurement with the 8 mm geometry made at intermediate/low temperatures. For Lab3, outlier data points were observed at high temperatures. With the 8 mm geometry, the curves follow the same smooth trend despite variability of the results, especially going forward with low shear modulus corresponding to higher temperature measurement. As compared to the neat bituminous binder, the curves were in the same range of magnitude. Figure 4b reports the measurement for the 25 mm made at high temperature regime above 33 °C. All curves displayed high scatter and no trends between laboratories can be identified. There was no smooth overlapping between temperatures, meaning the time temperature superposition can't be applied in this case.

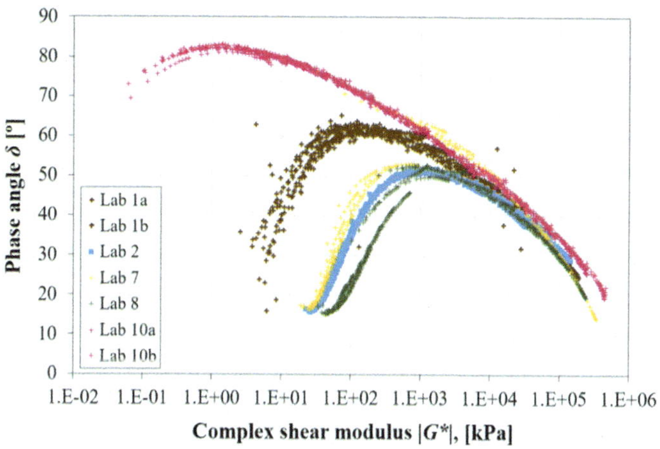

Fig. 3 Black space for $B_{+pellets}$

4.3 MSCR and LAS Tests Bituminous Binder

Figure 5 shows the MSCR results at 60 °C for both stress levels for binders B, $B_{+pellets}$, and $B_{+shreds}$. It can be observed that as the stress level increases from 0.1 to 3.2 kPa, the R decreases and the J_{nr} increases. Additionally, at both stress levels, the neat binder had the greatest J_{nr} and lowest R values. These findings indicate that, in comparison to mixes with PE shreds and pellets, the clean bituminous binder is more susceptible to persistent deformation. Additionally, because of its greater R and lower J_{nr} values, $B_{+shreds}$ routinely outperform $B_{+pellets}$ in terms of deformation recovery and permanent deformation sensitivity.

Figure 6 presents the fatigue results of the neat binder, B, the blend with the PE-pellets and PE-shreds, $B_{+pellets}$ and $B_{+shreds}$, at 20 °C. At the low strain level (lower than 2%), the results show that the neat binder has the lowest N_f values compared to the $B_{+shreds}$ and $B_{+pellets}$. It expresses that the binder containing PE shows lower fatigue life than virgin binder. On the other hand, the fatigue life of plastic modified binders ($B_{+shreds}$ and $B_{+pellets}$) was shorter than that of neat binder at high strain level (higher than 4%). These findings indicate that as bituminous binder becomes stiffer and brittle because of plastic modification, it can undertake more loading cycles under low strain levels; on the contrary, it can sustain less loading cycles under high strain levels due to the increase in brittleness. In addition, the $B_{+shreds}$ shows consistently better fatigue resistance $B_{+pellets}$ at any strain level.

The damage characteristics (C–Damage Intensity) curves of the tested binders are compared in Fig. 7. C denotes the ratio of the loss shear modulus to the starting value at any given time. By comparing the damage curves, it can be observed that the PEN 70/100 neat binder has the least damage compared with $B_{+shreds}$ and $B_{+pellets}$. When comparing the modified binders, it can be seen that $B_{+pellets}$ display somewhat less damage than $B_{+shreds}$.

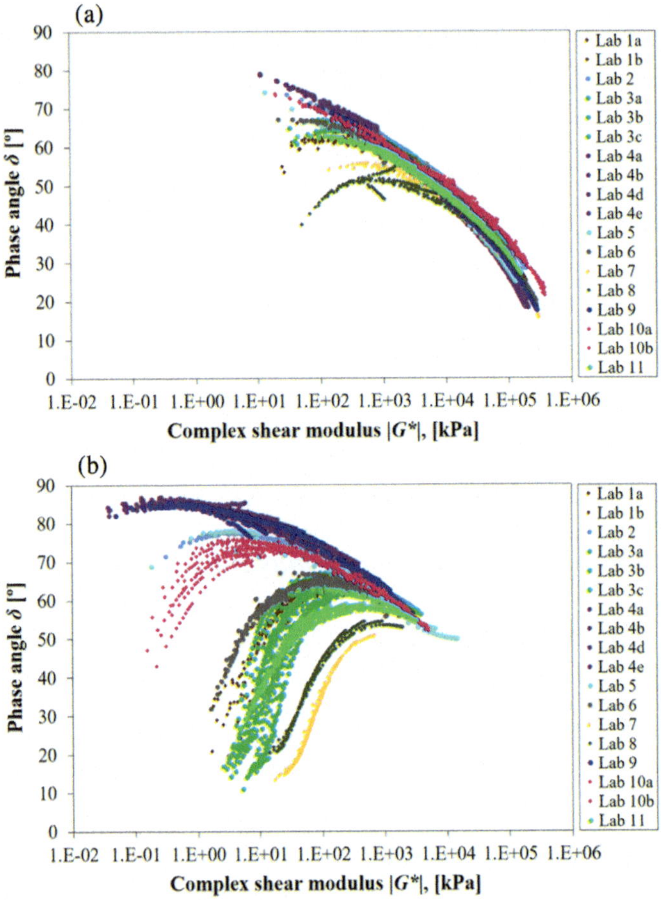

Fig. 4 Black diagram for $B_{+shreds}$ measured with: **a** 8 mm geometry; **b** 25 mm geometry [53]

4.4 FTIR and DSC Tests

Figure 8 displays the whole FTIR spectrum between 4000 and 600 cm^{-1} for the base bituminous binder 70/100, the blend with plastic shreds and the blend with pellets. Figure 9 provides a close up of most specific wavelength numbers for bituminous binder. From that spectrum it was not possible to identify any differences and as compared to the base bituminous binder 70/100 no additional peaks were detectable in the blends with plastics. This may have been an artifact of the measurement. The plastic having not been fully dispersed homogeneously in the bituminous binder matrix, there was only 5% of probability of having some plastic exactly under the diamond. Would have been the blend homogeneous continuum phase of plastic and bituminous binder, additional peaks should have been identified as the footprint of the plastic. An attend of FTIR on the plastic itself was also made but was not successful

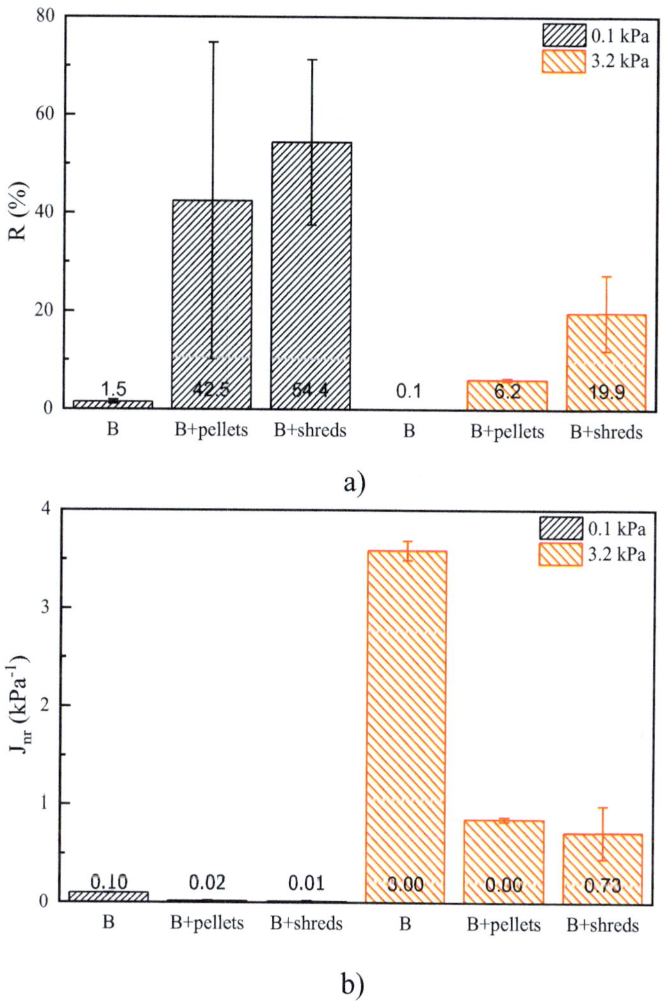

Fig. 5 **a** Strain recovery, and **b** non-recoverable creep compliance at two stress levels

due to the difficulty to apply the sample straight on the diamond ensuring proper contact for the measurement.

Differential scanning calorimetry was performed first on the plastic shreds and pellets even on several shreds samples differentiated by color. Figure 10 displays the heat flow during the heating phase from −60 to +165 °C. A single melting point was identified around 124 and 130 °C more pronounced for the pellets. No other outside the temperature range of the measurement.

Then DSC was performed on the base bituminous binder 70/100 and the two blends with shreds and pellets plastics. Figure 11 shows the heat flow during the heating phase for the three binders. The glass transition that can be observed for the

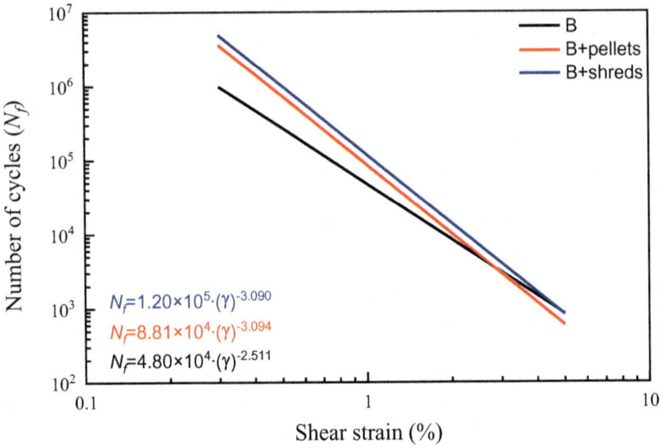

Fig. 6 Fatigue performances of three studied binders

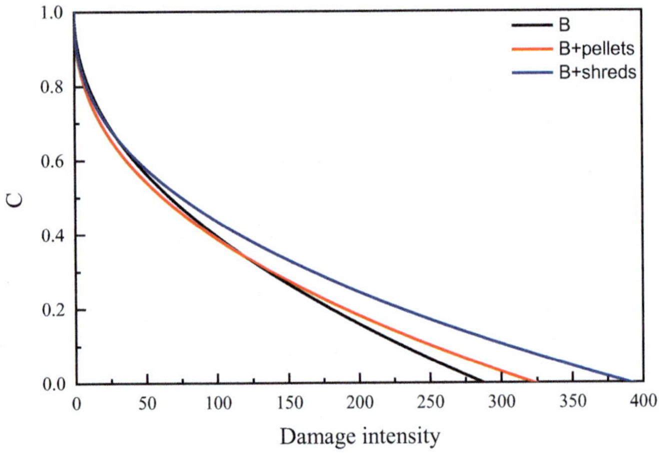

Fig. 7 C-Damage Intensity curves of three studied binders

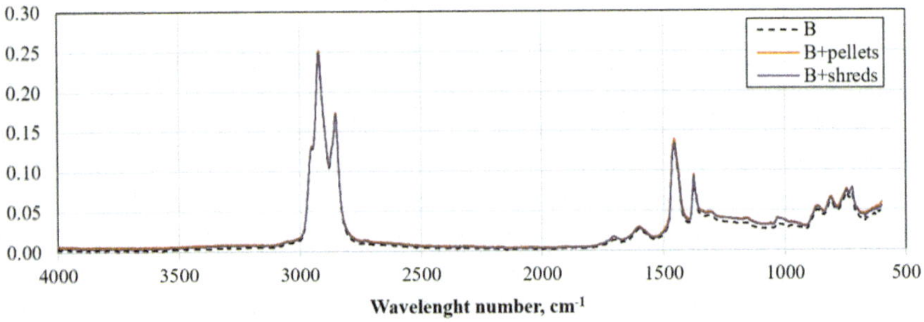

Fig. 8 FTIR spectrometry of 70/100 bituminous binder, blenda with shreds and pellets

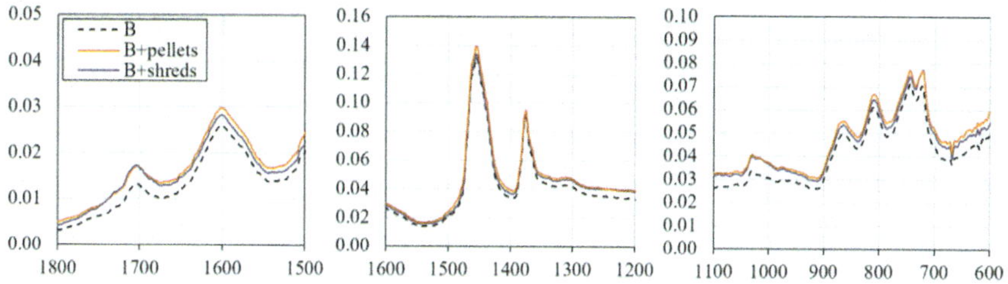

Fig. 9 Close up FTIR spectrometry

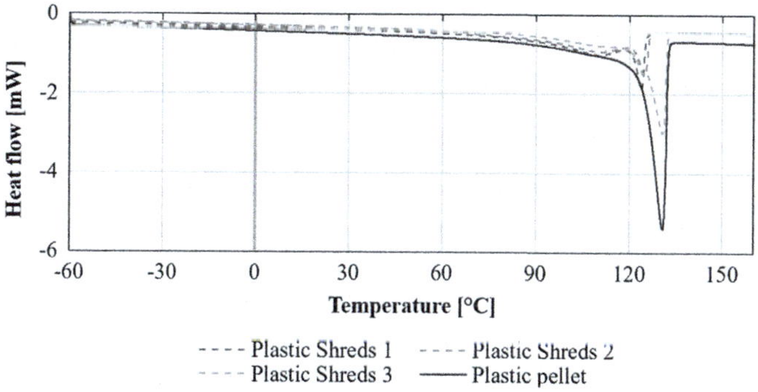

Fig. 10 Heat flow during heating process of plastic shreds and pellets

neat bituminous binder around −25 °C is also visible with the shreds and pellets blends. However, the latter clearly distinguishes the melting point of the plastic around 120 °C. Should have been the blend homogeneous with continuum phase morphology, the resulting glass transition would have been in between the base bituminous binder and the plastic. Both components maintained their specific calorimetric behaviour without any interaction, meaning the material has a two-phase morphology.

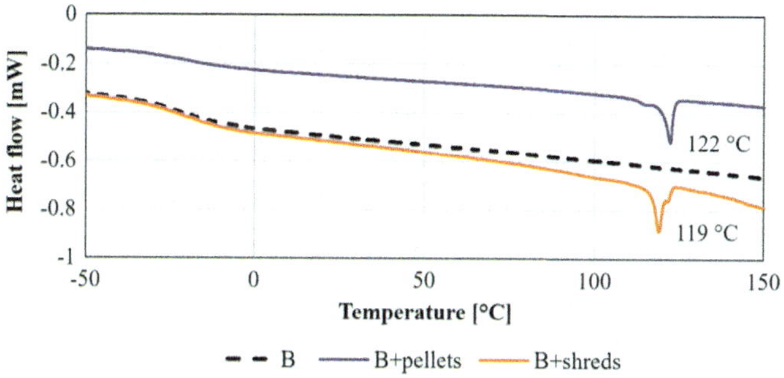

Fig. 11 Heat flow during heating process of 70/100 and plastic blends

The storage stability of the binder blends was assessed by one lab using the binder stability test [52]. These reveal that both PE blended binders show a significant difference in viscosity of the top and bottom samples and therefore, have a storage stability problem [49]. Softening point results not reliable, DSC can be a better tool to differentiate them.

5 Bituminous Mixture Modified with Waste PE

Given the promising results as well as the challenging issues observed at binder level and described in detail above, an interlaboratory exercise on bituminous mixtures modified with waste PE was carried out. This research activity was carried out by a selection of eleven laboratories from nine countries already involved in TG1 and/or TG3 (i.e., waste aggregates in bituminous mixtures) activities. This experimental study was aimed at determining if performance enhancement can be also observed at mixture level while limiting/hindering the abovementioned issues concerning stability and homogeneity.

To accomplish this objective, the selected laboratories were asked to prepare and characterize a dense graded bituminous mixture with 16 mm maximum aggregate size and containing different amounts of waste PE. In particular, each laboratory used its own aggregates and bituminous binder whereas waste PE was the same shreds used in the binder modification and was sampled and provided by a single source. Figure 12 shows particle size distribution of waste PE used by the different labs.

A common target mixture gradation (Fig. 13) as well as aggregate and bituminous binder type (i.e., limestone aggregates and 50/70 penetration grade bituminous

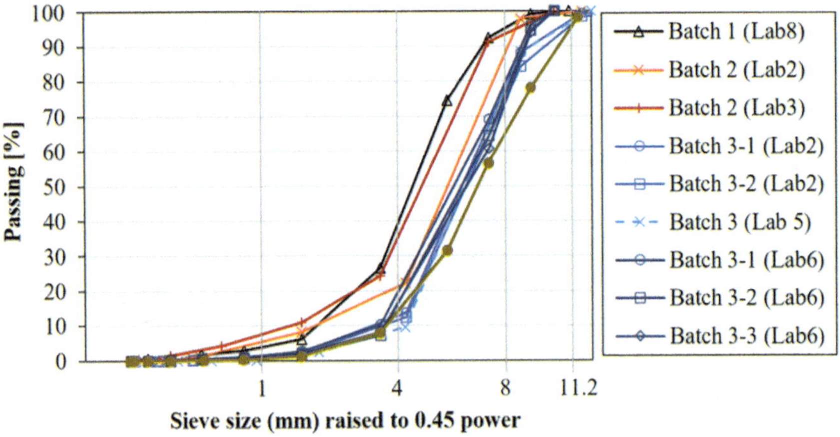

Fig. 12 Gradation of PE waste batches used by the different laboratories [59]

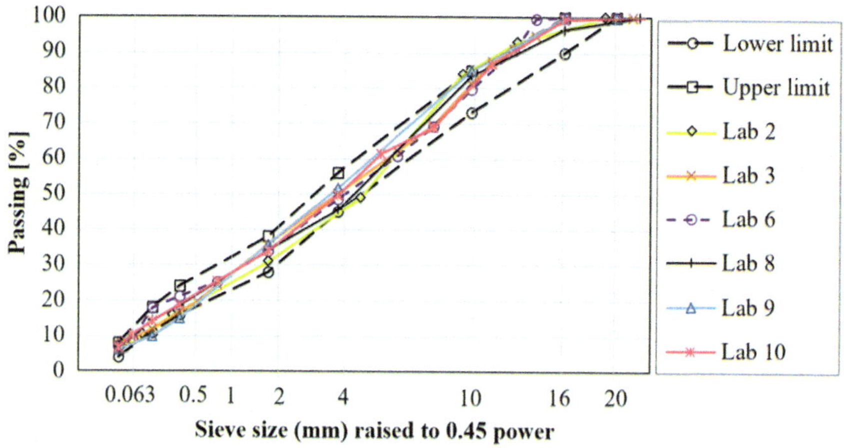

Fig. 13 Aggregate mix gradation adopted by participating labs [59]

binder, respectively) were suggested in order to minimize variations among laboratories. Only two laboratories used aggregates of different origin (i.e., granite for Lab2 and natural river sand for Lab6) while Lab8 used a 70/100 pen grade bituminous binder.

A binder content between 4% and 6% by aggregate weight was used in order to maximize mix properties and obtain a void content of 4%–6% when compacted with impact (i.e., Marshall) compactor or using a gyratory compactor. As far as the content of waste plastic concerns, all the laboratories investigated 1.5% plastic dosage by mix weight along with the reference materials (i.e., the mixture prepared without any plastic addition). Moreover, 0.25% (corresponding to 5.0% by bituminous binder weight as established for wet modification of binder discussed in the previous sections), 0.5%, 1.0% and 5.0% waste PE were also used by some selected laboratories as detailed in Table 2.

The mixtures were prepared in the laboratory using a protocol which can be considered as a dry process since the plastic at ambient temperature was added to the hot aggregates (160 °C) and premixed for few minutes (1–7 min depending on the batch size) prior to the addition of the hot bituminous binder. This allowed to achieve a homogeneous mix without a complete melting of the plastic particles. Cylindrical or slab specimens were then compacted at 155 °C using Marshall, gyratory or roller compactor compliant with EN standards.

The reference and PE modified bituminous mixes were characterized in terms of workability, tensile strength, stiffness, permanent deformation as well as moisture resistance. Workability was assessed by analyzing the gyratory compaction curves while the indirect tensile strength test carried out in both dry and wet conditions was used to evaluate the strength and the moisture resistance, respectively. On the other hand, stiffness and linear visco-elastic characteristics were studied through pulse indirect tension tests and cyclic tension–compression tests at intermediate service

Table 2 Bituminous mixture properties

PE %	Lab1	Lab2	Lab3	Lab4	Lab5	Lab6	Lab7	Lab8	Lab9	Lab10	Lab11
	Void content (vol % in asphalt sample)										
0	5.30	3.60	3.54	5.60	4.30	5.54	6.59	4.50	6.00	3.70	5.54
0.25	5.10	4.00	3.48					4.50	5.80		
0.5											5.14
1.0											5.05
1.5	4.80	5.00	5.93	4.40	4.50	4.59	3.13	4.50	4.70	5.70	
5					5.47				2.70		
	Bituminous binder content (wt% by weight of stone aggregate)										
0	4.00	4.70	4.10	5.00	5.00	4.71	5.4	5.00	5.30	4.88	5.26
0.25	4.00	4.70	4.10					5.00	5.30		
0.5											5.29
1.0											5.32
1.5	4.00	4.70	4.10	5.00	5.00	4.71	5.4	5.00	5.30	4.90	
5					5.00				5.30		

temperatures (25–35 °C). Finally, cyclic compression tests (CCT) on cylindrical specimens or wheel tracking tests (WTT) on slabs were used to analyze the resistance of the investigated materials against permanent deformation at 60 °C. More details about materials, specimen preparation and testing methods are reported elsewhere [60].

The experimental findings achieved at multi-laboratory level mainly showed that the presence of waste PE did not negatively affect mix workability. In particular, lower construction densification index (CDI) and compactability parameter *K* were generally found for plastic mixtures as shown in the following Figs. 14 and 15. The error bars indicate the standard deviation of the results in Fig. 14 and the following plots.

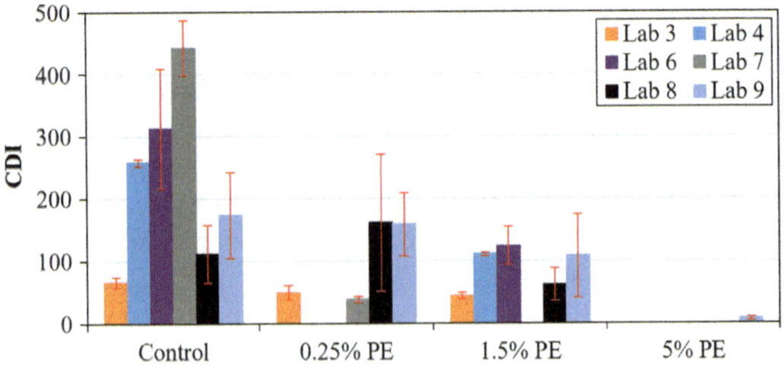

Fig. 14 CDI of the investigated mixtures [60]

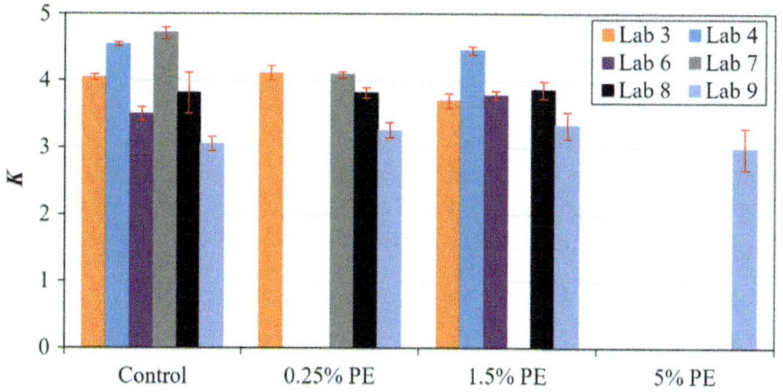

Fig. 15 Compactability *K* of the investigated mixtures [60]

Such experimental findings suggest that the investigated amounts of waste plastics would not lead to workability issues thus allowing a proper laying of such materials. Furthermore, enhanced elasticity, stiffness and permanent deformation resistance were also observed by the different laboratories.

In this sense, Fig. 16 depicts the higher strength achieved thanks to the inclusion of waste PE; overall, the higher the PE dosage the higher the ITS increase with respect to the values measured for the control materials (i.e., without any plastic waste) likely thanks to the improved cohesion of the plastic modified bituminous mastic. It is worth noting that average absolute ITS values varied between 1.09 and 1.58 MPa.

The dynamic tests at intermediate temperatures mostly confirmed the performance enhancement achievable with the inclusion of the selected waste PE shreds into the bituminous mixtures since an increase in stiffness, a lower time dependence and a higher elasticity were often measured when plastic was added (Figs. 17 and 18).

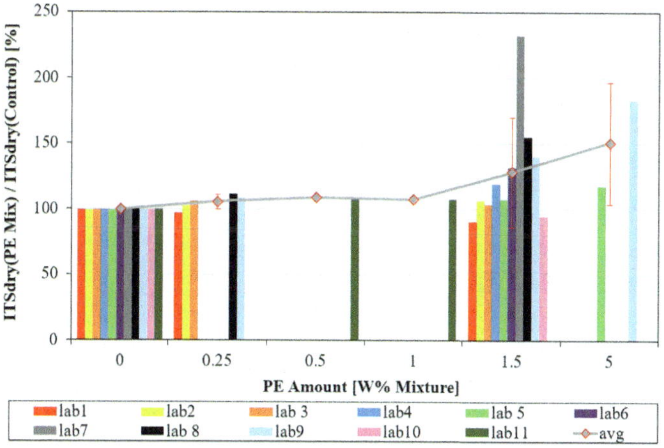

Fig. 16 ITS increase as a function of waste PE content [59]

Furthermore, a higher variability among the different labs was noticed likely due to a higher sensitivity of such properties to the air void content of the tested specimens.

The enhanced stiffness and elasticity of the plastic mixes reflected in a clear improvement of their permanent deformation resistance. In this regard, CCT tests demonstrated a reduced creep rate along with a higher creep modulus thanks to the addition of PE shreds as shown in Fig. 19; similar conclusions can be drawn from the experimental findings coming from WTS (Fig. 20).

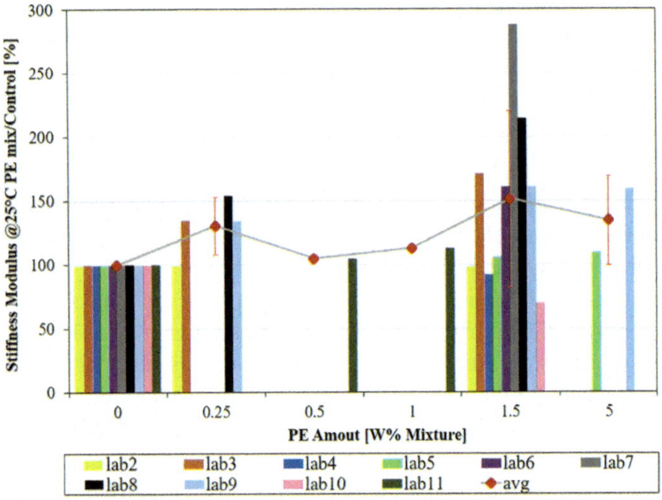

Fig. 17 Stiffness characteristics as a function of waste PE content [59]

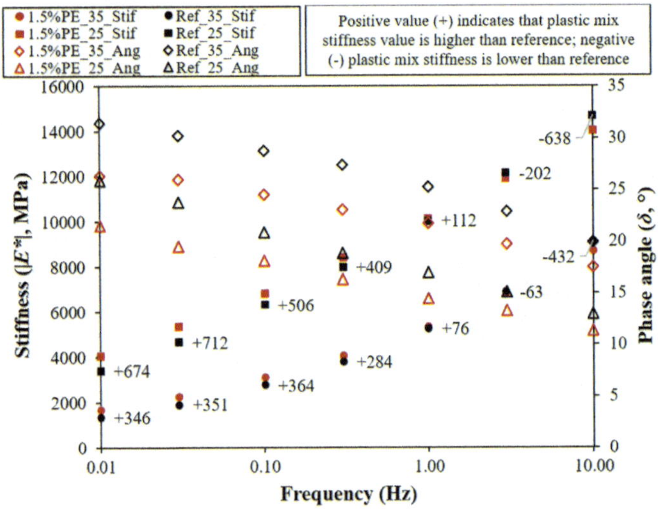

Fig. 18 Linear visco-elastic properties as a function of waste PE content [59]

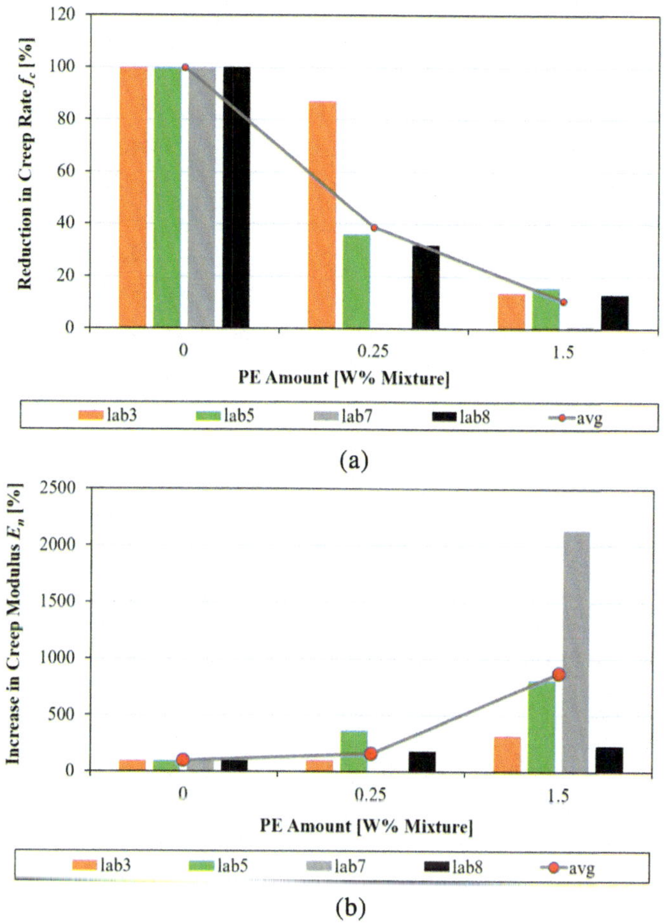

Fig. 19 **a** Change in creep rate, and **b** increase in creep modulus as a function of waste PE content [60]

Finally, acceptable (often even improved) moisture resistance was also observed for PE modified materials. Indeed, the percentage ITS ratio measured in wet and dry conditions (ITSR) is reported in Fig. 21 as a function of the plastic content. In Fig. 21, it is worth noting that ITSR absolute values higher than 80% (i.e., value accepted by the most part of highway agencies worldwide) were measured by all the involved laboratories, regardless of the presence of PE waste plastic. The only exception was Lab2 which tested bituminous mixtures prepared with water sensitive granite aggregates.

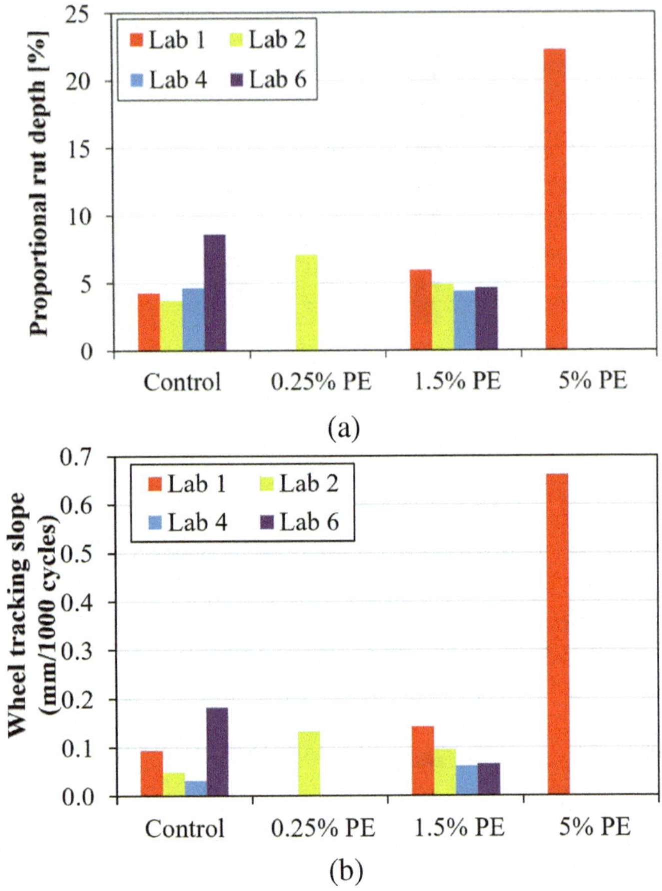

Fig. 20 **a** Proportional rut depth (PRD), and **b** wheel tracking slope (WTS) as a function of waste PE content [60]

6 Summary and Conclusions

The research work performed by TG1 was presented and summarized in this second chapter of the State-of-the-Art Report of the RILEM TC 279-WMR. This investigation was devoted to the combination of waste Polyethylene (PE) plastic in bituminous binder and asphalt mixture due to the characteristics of these materials. Two forms of PE were evaluated: pellets and shreds obtained from packaging and its recycling in the form of pellets and the secondary waste produced as a result, respectively. The research conducted on bituminous binder modified with PE consisted of a comprehensive experimental program supported by an advanced data analysis. The program was performed on plain binder, binders modified with PE shreds and PE pellets, and included conventional characterization, rheological testing with the Dynamic Shear

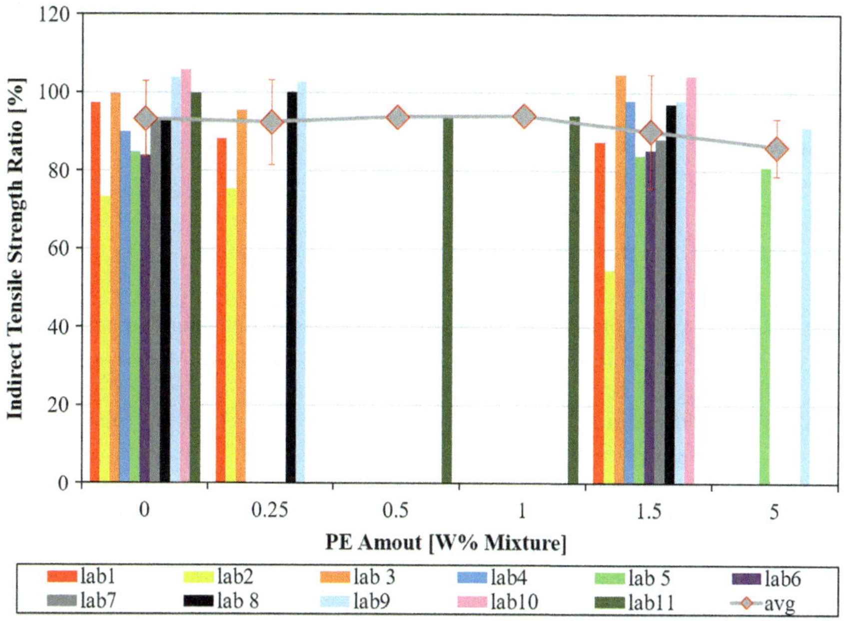

Fig. 21 ITSR as a function of waste PE content [59]

Rheometer (DSR), Differential Scanning Calorimetry (DSC), and Fourier Transform Infrared Spectroscopy (FTIR) for chemical analysis. PE-modified bituminous mixtures were also prepared to expand the understanding of the effect of PE on paving mixtures. The bituminous mixture was designed according to the dry process, and a series of mechanical tests were conducted to determine stiffness, strength, strength ratio, and resistance to permanent deformations. Based on the experimentation and analysis performed, the following conclusions can be drawn:

- At high temperatures, using the Multiple Stress Creep Recovery (MSCR) test, the binder blends modified with PE were less sensitive to permanent deformation compared to the non modified binder.
- At intermediate temperatures using the Linear Amplitude Sweep (LAS) tests, the fatigue performance of the PE blends could withstand more loading cycles under low strain levels; however, it could sustain less loading cycles under high strain levels due to the increase in brittleness.
- The addition of waste PE did not significantly affect the workability of asphalt mixtures.
- The higher the PE dosage, the higher the Indirect Tensile Strength (ITS) increase with respect to the values measured for the control materials (i.e., without any plastic waste) thanks to the improved cohesion of the plastic modified mastic.
- The stiffness experiments showed an improved performance with a lower time dependence and a higher elasticity when plastic was added.

- The cyclic compression tests demonstrated a reduced creep rate along with a higher creep modulus thanks to the addition of PE shreds; similar conclusions can be drawn from the experimental findings from the wheel tracking test.
- Acceptable and often improved moisture resistance was observed for PE modified mixtures.

7 Perspective and Outlook

The activities of TG1 of TC-297 WMR summarized in this chapter demonstrate that the use of PE in bituminous mixtures can be a viable option. However, appropriate test methods need to be used to evaluate the performance of the material. As binder modified with PE specimens can have the disadvantage of segregation and inhomogeneity, mixture performance should be preferably used as the deciding procedure. A balanced mix design that considers the low temperature and high temperature performance should be considered. In addition, environmental issues should be part of the evaluation procedure such as, for example, leaching risks. In order to determine the viability of using waste materials throughout their life cycle, a complete life cycle assessment preferably a cradle to grave procedure should be considered. In such analysis the waste management procedure in different geographical areas should be considered as using waste in bituminous mixture can take this material away from another industry (e.g., cement industry where plastic waste is used as fuel [61]) and this would have consequences for that industry that could not be necessarily sustainable.

References

1. European Commission. Waste management hierarchy. https://ec.europa.eu/environment/green-growth/waste-prevention-and-management/index_en.html. Accessed 21 Nov 2021
2. Lo Presti D (2013) Recycled tyre rubber modified bitumens for road asphalt mixtures: a literature review. Constr Build Mater 49:863–881. https://doi.org/10.1016/j.conbuildmat.2013.09.007
3. Piao Z, Mikhailenko P, Kakar MR, Bueno M, Hellweg S, Poulikakos LD (2021) Urban mining for asphalt pavements: a review. J Clean Prod 280:124916. https://doi.org/10.1016/j.jclepro.2020.124916
4. Poulikakos LD, Papadaskalopoulou C, Hofko B, Gschösser F, Cannone Falchetto A, Bueno M, Arraigada M, Sousa J, Ruiz R, Petit C, Loizidou M, Partl MN (2017) Harvesting the unexplored potential of European waste materials for road construction. Resour Conserv Recycl 116:32–44. https://doi.org/10.1016/j.resconrec.2016.09.008
5. AASHTO T315 (2022) Standard method of test for determining the rheological properties of asphalt binder using a dynamic shear rheometer (DSR). American Association of State Highway and Transportation Officials
6. Farrar M, Sui C, Salmans S, Qin Q (2015) Determining the low-temperature rheological properties of asphalt binder using a dynamic shear rheometer (DSR). Technical white paper FP08. No. DTFH61-07-D-00005, Fundamental properties of asphalts and modified asphalts, III

7. Wang D, Cannone Falchetto A, Alisov A, Schrader J, Riccardi C, Wistuba MP (2019) An alternative experimental method for measuring the low temperature rheological properties of asphalt binder by using 4mm parallel plates on dynamic shear rheometer. Transp Res Rec 2673(3):427–438. https://doi.org/10.1177/0361198119834 91
8. Moon KH, Cannone Falchetto A, Wang D, Wistuba MP, Tebaldi G (2017) Low-temperature performance of recycled asphalt mixtures under static and oscillatory loading. Road Mater Pavement Des 18(2):297–314. https://doi.org/10.1080/14680629.2016.1213500
9. Leaderman H (1943) Elastic and creep properties of filamentous materials and other high polymers. Textile Foundation, Washington
10. Williams ML, Landel RF, Ferry JD (1955) The temperature dependence of relaxation mechanisms in amorphous polymers and other glass-forming liquids. J Am Chem Soc 77(14):3701–3707. https://doi.org/10.1021/ja01619a008
11. Porot L, Eduard P (2016) Addressing asphalt binder aging through the viscous to elastic transition. In: Proceedings of the ISAP symposium 2016, Jackson Hole Wyoming, USA
12. Erkens S, Porot L, Glaser R, Glover C (2016) Review of asphalt (concrete) aging tests in US and Europe. In: The 95th Transportation Research Board annual meeting, Washington D.C, 16-5770
13. Glover CJ, Davison RR, Domke CH, Ruan Y, Juristyarini P, Knorr DB, Jung SH (2005) Development of a new method for assessing asphalt binder durability with field validation. FHWA/TX-05/1872-2, Federal Highway Administration and Texas Department of Transportation, College Station, TX, 1872, pp 1–334
14. Rowe GM (2011) Prepared discussion following the Anderson AAPT paper cited previously. In: Association of Asphalt Paving Technologists, vol 80, pp 649–662
15. Rowe GM (2014) Interrelationships in rheology for asphalt binder specifications. In: Proceedings of the fifty-ninth annual conference of the Canadian technical asphalt association (CTAA), Winnipeg, Manitoba
16. Cannone Falchetto A, Moon KH, Wang D, Riccardi C (2018) Investigation on the cooling medium effect in the characterization of asphalt binder with the bending beam rheometer (BBR). Can J Civ Eng 45(7):594–604. https://doi.org/10.1139/cjce-2017-0586
17. AASHTO T313 (2012) Standard method of test for determining the flexural creep stiffness of asphalt binder using the bending beam rheometer (BBR). American Association of State Highway and Transportation Officials
18. Nasr D, Pakshir AH (2019) Rheology and storage stability of modified binders with waste polymers composites. Road Mater Pavement Des 20(4):773–792. https://doi.org/10.1080/14680629.2017.1417152
19. Kumar P, Garg R (2011) Rheology of waste plastic fibre-modified bitumen. Int J Pavement Eng 12(5):449–459. https://doi.org/10.1080/10298430903255296
20. Sadeque M, Patil KA (2013) Rheological properties of recycled low density polyethylene and polypropylene modified bitumen. Int J Civ Eng 2(2):24–26. https://doi.org/10.47893/IJATCE.2013.1076
21. Singhal M, Yadav Y, Mandal R (2016) Use of modified bituminous binder in highway construction. Int J Innov Res Technol Sci Eng 2:176–382
22. Hu C, Lin W, Partl M, Wang D, Yu H, Zhang Z (2018) Waste packaging tape as a novel bitumen modifier for hot-mix asphalt. Constr Build Mater 193:23–31. https://doi.org/10.1016/j.conbuildmat.2018.10.170
23. Tušar M, Kakar MR, Poulikakos LD, Pasquini E, Baliello A, Pasetto M, Porot L, Wang D, Cannone Falchetto A, Dalmazzo D, Lo Presti D, Giancontieri G, Varveri A, Veropalumbo R, Viscione N, Vasconcelos K, Carter A (2022) RILEM TC 279 WMR round robin study on waste polyethylene modified bituminous binders: advantages and challenges. Road Mater Pavement Des 1–29. https://doi.org/10.1080/14680629.2021.2017330
24. D'Angelo JA (2009) The relationship of the MSCR test to rutting. Road Mater Pavement Des 10:61–80. https://doi.org/10.1080/14680629.2009.9690236
25. Zhang J, Faruk AN, Karki P, Holleran I, Hu X, Walubita LF (2016) Relating asphalt binder elastic recovery properties to HMA cracking and fracture properties. Constr Build Mater 121:236–245. https://doi.org/10.1016/j.conbuildmat.2016.05.157

26. Chen S, Che T, Mohseni A, Azari H, Heiden PA, You Z (2021) Preliminary study of modified asphalt binders with thermoplastics: the rheology properties and interfacial adhesion between thermoplastics and asphalt binder. Constr Build Mater 301:124373. https://doi.org/10.1016/j.conbuildmat.2021.124373
27. Joohari I, Giustozzi F (2022) Oscillatory shear rheometry of hybrid polymer-modified bitumen using multiple stress creep and recovery and linear amplitude sweep tests. Constr Build Mater 315:125791. https://doi.org/10.1016/j.conbuildmat.2021.125791
28. Delgado-Jojoa MG, Sánchez-Gilede JA, Rondón-Quintana HA, Fernández-Gómez WD, Reyes-Lizcano FA (2018) Influence of four non-conventional additives on the physical, rheological and thermal properties of an asphalt. Ing e Investig 38(2):18–26. https://doi.org/10.15446/ing.investig.v38n2.68638
29. Nuñez JYM, Domingos MDI, Faxina AL (2014) Susceptibility of low-density polyethylene and polyphosphoric acid-modified asphalt binders to rutting and fatigue cracking. Constr Build Mater 73:509–514. https://doi.org/10.1016/j.conbuildmat.2014.10.002
30. Zhou HY, Dou HB, Chen XH (2021) Rheological properties of graphene/polyethylene composite modified asphalt binder. Materials 14(14):1–15. https://doi.org/10.3390/ma14143986
31. ASTM D2872-04 (2004) Standard test method for effect of heat and air on a moving film of asphalt (Rolling Thin-Film Oven Test). ASTM International, West Conshohocken, PA, USA
32. EN 12607-1 (2014) Bitumen and bituminous binders—determination of the resistance to hardening under influence of heat and air—part 1: RTFOT method
33. ASTM D6521-08 (2008) Standard practice for accelerated aging of asphalt binder using a pressurized aging vessel (PAV). ASTM International, West Conshohocken, PA, USA
34. EN 14769 (2012) Bitumen and bituminous binders—accelerated long-term ageing conditioning by a pressure ageing vessel (PAV). European Committee for Standardization, Brussels, Belgium
35. Soenen H, Besamusca J, Fischer HR, Poulikakos LD, Planche JP, Das PK et al (2014) Laboratory investigation of bitumen based on round robin DSC and AFM tests. Mater Struct 47(7):1205–1220. https://doi.org/10.1617/s11527-013-0123-4
36. ISO 11357-1 (2009) Plastics-differential scanning calorimetry, differential scanning calorimetry. International Organization for Standardization
37. ASTM 1356-08 (2014) Standard test method for assignment of the glass transition temperatures by differential scanning calorimetry. ASTM International, West Conshohocken, PA, USA
38. ASTM D 3418-21 (2021) Standard test method for transition temperatures and enthalpies of fusion and crystallization of polymers by differential scanning calorimetry. ASTM International, West Conshohocken, PA, USA
39. Masson JF, Polomark GM (2001) Bitumen microstructure by modulated differential scanning calorimetry. Thermochim Acta 374(2):105–114. https://doi.org/10.1016/S0040-6031(01)00478-6
40. Petersen JC (1986) Quantitative functional group analysis of asphalts using differential infrared spectrometry and selective chemical reactions–theory and application. Transp Res Rec 1096:1–11
41. Hofko B, Porot L, Cannone Falchetto A, Poulikakos LD, Huber L, Lu X et al (2018) FTIR spectral analysis of bituminous binders: reproducibility and impact of ageing temperature. Mater Struct 51(2):1–16. https://doi.org/10.1617/s11527-018-1170-7
42. Lamontagne J, Dumas P, Mouillet V, Kister J (2001) Comparison by Fourier transform infrared (FTIR) spectroscopy of different ageing techniques: application to road bitumens. Fuel 80(4):483–488. https://doi.org/10.1016/S0016-2361(00)00121-6
43. Mouillet V, Lamontagne J, Durrieu F, Planche J, Lapalu L (2008) Infrared microscopy investigation of oxidation and phase evolution in bituminous binder modified with polymers. Fuel 87(7):1270–1280. https://doi.org/10.1016/j.fuel.2007.06.029
44. Porot L, Mouillet V, Margaritis A, Haghshenas H, Elwardany M, Apostolidis P (2022) Fourier-transform infrared analysis and interpretation for bituminous binders. Road Mater Pavement Des 1–22. https://doi.org/10.1080/14680629.2021.2020681

45. EN 12697-10 (2017) Bituminous mixtures—test methods—part 10: compactability. European Committee for Standardization, Brussels, Belgium
46. EN 12697-23 (2017) Bituminous mixtures—test methods—part 23: determination of the indirect tensile strength of bituminous specimens. European Committee for Standardization, Brussels, Belgium
47. EN 12697-25 (2016) Bituminous mixtures—test methods—part 25: cyclic compression test. European Committee for Standardization, Brussels, Belgium
48. EN 12697-22 (2020) Bituminous mixtures—test methods—part 22: wheel tracking. European Committee for Standardization, Brussels, Belgium
49. Kakar MR, Mikhailenko P, Piao Z, Bueno M, Poulikakos L (2021) Analysis of waste polyethylene (PE) and its by-products in asphalt binder. Constr Build Mater 280:122492. https://doi.org/10.1016/j.conbuildmat.2021.122492
50. EN 1426 (2015) Bitumen and bituminous binders—determination of needle penetration. European Committee for Standardization, Brussels, Belgium
51. EN 1427 (2015) Bitumen and bituminous binders—determination of the softening point—ring and ball method. European Committee for Standardization, Brussels, Belgium
52. EN 13399 (2017) Bitumen and bituminous binders—determination of storage stability of modified bitumen. European Committee for Standardization, Brussels, Belgium
53. Wang D, Baliello A, Poulikakos LD, Vasconcelos K, Pinheiro G, Kakar MR, Giancontieri G, Pasquini E, Porot L, Tušar M, Riccardi C, Pasetto M, Lo Presti D, Cannone Falchetto A (2022) Rheological properties of asphalt binder modified with waste polyethylene: an interlaboratory research from the RILEM TC WMR. Resour Conserv Recycl 186:106564. https://doi.org/10.1016/j.resconrec.2022.106564
54. Wang D, Baliello A, Pinheiro G, Poulikakos LD, Tušar M, Vasconcelos K, Kakar MR, Porot L, Pasquini E, Giancontieri G, Riccardi C, Pasetto M, Lo Presti D, Cannone Falchetto A (2022) Rheological behaviors of waste polyethylene modified asphalt binder: statistical analysis of inter-laboratory testing results. J Test Eval 51(4):1–2. https://doi.org/10.1520/JTE20220313
55. AASHTO TP 70 (2013) Standard method of test for multiple stress creep recovery (MSCR) test of asphalt binder using a dynamic shear rheometer (DSR). American Association of State Highway and Transportation Officials
56. EN 16659 (2015) Bituminen and bituminous binders—multiple stress creep and recovery test (MSCRT). European Committee for Standardization, Brussels, Belgium
57. AASTHO TP 101-12 (2012) UL standard method of test for estimating fatigue resistance of asphalt binders using the linear amplitude sweep. American Association of State and Highway Transportation Officials
58. Schapery RA, Park SW (1999) Methods of interconversion between linear viscoelastic material functions. Part II—an approximate analytical method. Int J Solids Struct 36(11):1677–1699. https://doi.org/10.1016/S0020-7683(98)00055-9
59. Johnson C (2010) Estimating asphalt binder fatigue resistance using an accelerated tested method. PhD thesis, University of Wisconsin-Madison, Madison, USA. http://digital.library.wisc.edu/1793/46799
60. Poulikakos LD, Pasquini E, Tušar M, Hernando D, Wang D, Mikhailenko P, Pasetto M, Baliello A, Stoop J, Wouters L, Cannone Falchetto A, Miljković M, Orešković M, Viscione N, Saboo N, Orozco G, Lachance-Tremblay É, Vaillancourtj M, Kakar MR, Bueche N, Dalmazzo D, Pinheiro G, Vasconcelos K, Moreno Navarro F (2022) RILEM interlaboratory study on the mechanical properties of asphalt mixtures modified with polyethylene waste. J Clean Prod 375:133124. https://doi.org/10.1016/j.jclepro.2022.133124
61. Piao Z, Bueno M, Poulikakos LD, Hellweg S (2022) Life cycle assessment of rubberized semi-dense asphalt pavements; a hybrid comparative approach. Resour Conserv Recycl 176:105950. https://doi.org/10.1016/j.resconrec.2021.105950

Chapter 3
Crumb Rubber Modified Binders

Jorge C. Pais, Lily D. Poulikakos, Patricia Kara De Maeijer, Nicolas Schüwer, Maria Chiara Cavalli, Augusto Cannone Falchetto, Muhammad Rafiq Kakar, Johan Blom, Maeva Tobler, Marcel Perecmanis, Di Wang, and Fucheng Guo

Abstract RILEM Technical Committee 279 WMR is dedicated to the Valorization of Waste and Secondary Materials for Roads. Its Task Group 2 investigated Crumb Rubber (CR) as an additive to enhance the performance of bitumen. CR recycled from end-of-life tires (ELTs) was chosen for this investigation because crumb rubber modified bitumen (CRMB) has been used to improve bituminous mixtures performance for fatigue and reflective cracking. The success of these mixtures is due to the CRMB viscosity that allows the use of an increased amount of bitumen compared to conventional mixtures. Because the viscosity of the CRMB is a function of the CR surface, and presently various types of CRs are produced, it is crucial to verify how these materials perform as a bitumen modifier. Interlaboratory experiments were performed on four types of CR, obtained from mechanical grinding, cryogenic process, waterjet pulverization and reacted and activated rubber. Three

J. C. Pais (✉) · M. Perecmanis
ISISE, Department of Civil Engineering, University of Minho, 4800-058 Guimarães, Portugal
e-mail: jpais@civil.uminho.pt

L. D. Poulikakos
EMPA, Swiss Federal Laboratories for Materials Science and Technology, Überlandstrasse 129, CH-8600 Dübendorf, Switzerland

P. K. De Maeijer · J. Blom
Faculty of Applied Engineering—EMIB, University of Antwerp, 171 Groenenborgerlaan 171, 2020 Antwerp, Belgium

N. Schüwer · M. Tobler
Tyre Recycling Solutions SA, Z.I. Trési 9A, 1028 Préverenges, Switzerland

M. C. Cavalli
School of Architecture and the Built Environment, KTH Royal Institute of Technology, Brinellvägen 23, 10044 Stockholm, Sweden

A. Cannone Falchetto · D. Wang · F. Guo
Department of Civil Engineering, Aalto University, Rakentajanaukio 4, 02150 Espoo, Finland

M. R. Kakar
Department of Architecture, Wood and Civil Engineering, Bern University of Applied Sciences (BFH), Pestalozzistrasse 20, 3400 Burgdorf, Switzerland

© The Author(s), under exclusive license to Springer Nature Switzerland AG 2023

A. Cannone Falchetto et al. (eds.), *Valorisation of Waste and Secondary Materials for Roads*, RILEM State-of-the-Art Reports 38,
https://doi.org/10.1007/978-3-031-33173-2_3

base, 35/50, 50/70 and 70/100, bitumen were used for the modification. Mechanical and chemical properties of CRMB were investigated. Despite some differences in the non-mechanical tests, i.e., penetration, softening point and viscosity, the results of the mechanical tests (complex shear modulus) suggest that the bitumen penetration grade ultimately dictates CRMB response.

Keywords Crumb rubber (CR) · Crumb rubber modified bitumen (CRMB) · FTIR · Viscosity · Rheology

1 Introduction

The use of waste and marginal materials in roads can contribute to the reduction of the use of primary materials as well as waste reduction. Assuming that using waste materials on roads results in similar long-term performance, this could provide a viable solution for producing sustainable road materials [1, 2].

RILEM Technical Committee 279 WMR on Valorisation of Waste and Secondary Materials for Roads (www.rilem.net) intended to achieve the following goals: (i) identify waste and marginal materials that are performance-enhancing components for road materials; (ii) investigate suitable bitumen additives to enhance the performance of the bitumen, suitable aggregate substitutes to enhance the performance of mixtures; (iii) explore possible polluting consequences; (iv) using Life Cycle Assessment (LCA) and Life Cycle Costs Analysis (LCCA) to show that these materials are environmentally and economically desirable; (v) recommending suitable waste materials and limit amounts for use in roads. The present chapter provides an overview of the research activity of Task Group 2 (TG2).

TG2 focused on studying the bitumen modification using crumb rubber (CR) from waste tires, considering the effect of different CR types on bitumen performance. Currently, the Technical Readiness Level (TRL) regarding the use of CR from waste tires in roads is very high [1], which poses the question of why this waste material is not used more widely worldwide. Perhaps this can be attributed to the lack of trust in the material performance and/or know-how on how to produce the material, where the results of this technical committee can contribute.

2 State-of-the-Art Review

2.1 End-of-Life Tires

The treatment, processing, and comprehensive utilization of end-of-life tires (ELTs) and rubber is critical for up-to-date waste management. By 2030, it is expected that ELTs would weigh a total of at least 1.2 billion tons. Only 30% of these ELTs are thought to be exported, recycled and reused, with the remaining 70% ending up in stockpiles, landfills, unlawful disposal, or other unreported places [3].

Even if avoidance and reduction are now impractical options, ELT reuse and recycling are unquestionably viable alternatives that should be chosen over energy recovery and disposal products as alternative building materials in sustainable large-scale civil engineering projects [4].

The most popular methods for reducing the environmental effects of end-of-life tires include recycling, reusing, retreading, re-generation, co-processing, pyrolysis, and landfills [5]. Research from the past supports the Waste Framework Directive's 2008/98/EC recommendations that, whenever possible, material recovery should take precedence over energy recovery [6].

2.2 *Crumb Rubber Production*

The use of CR with bitumen has a long history that began in the 1960s [7]. Presently, CR is obtained by mechanical grinding at ambient temperature, cryogenic milling, and waterjet pulverization. Mechanical grinding is the least expensive and most common method, and it relies on a multistage grinding and shredding process. In cryogenic milling, tire chips are frozen in liquid nitrogen and then size-reduced with a hammer mill. Waterjet pulverization employs ultra-high-water pressure to micronize tire rubber. Some processes, like waterjet technology, allow for specific processing from selected areas of the tire. Mechanical grinding typically uses whole tires, leading to powder composed of a mixture of rubber compounds.

Consequently, CR produced from those three processes usually presents different chemical compositions. The type of micronization process affects the chemical composition, but it also influences the rubber particle morphology. The waterjet pulverization yields rubber powder particles with a rough surface instead of the smooth and angular-shaped particles obtained from the cryogenic process [8, 9]. These morphological differences produce different CR surfaces. Previously reported values are between 0.044 and 0.064 m^2/g for mechanically ground crumbs; 0.031 and 0.044 m^2/g for cryogenically ground rubber powder and 0.115 m^2/g for waterjet rubber powder [10].

Reacted and Activated Rubber (RAR), a processed CR product, has been utilized to improve the stability, drain down, rutting, and fatigue resistance of bituminous mixtures. RAR is a short-term hot activation and short-time hot blending elastomeric bitumen extender created in a specially engineered technique to create dried granular activated rubber. According to [11, 12], RAR is made up of basic bitumen, finely crushed CR, and mineral filler.

The engineered CR technology is yet another type of activated rubber that has demonstrated performance improvement as compared to regular rubber powder as it contains filler and additives [13].

2.3 Crumb Rubber Modified Bitumen

The bitumen modification with CR powder has been subjected to abundant research efforts over the past decades. However, results from previously published research are difficult to compare to one another since different studies used different materials (type of bitumen and CR) and variable mixing conditions (temperature, time, or shearing rate, amongst others). Mixing variables are critical since process variables affect crumb rubber modified bitumen (CRMB) and bituminous mixture properties [14]. Nevertheless, CR typically increases the bitumen softening point and decreases its penetration index. These properties usually change linearly with the CR content [15]. Adding CR to bitumen further affects ductility, Fraass breaking point, and elastic recovery [16]. Dynamic Shear Rheometer (DSR) measurement of CRMB showed that CR increases the bitumen viscoelastic modulus and viscosity at high temperatures but, on the other hand, reduces the storage and loss modulus at low temperatures [17].

The changed bitumen viscosity can be impacted by the CR size [18], higher viscosity was achieved by using larger CR particle sizes [19], which ultimately increased deformation resistance [20]. Furthermore, low particle surface area reduces the bitumen's absorption process and enhances the rubber's digestion into the bitumen [21]. Therefore, it is strongly suggested to utilize recycled CR with particle sizes smaller than 1 mm. Compared to introducing bigger particles, it improved stiffness and led to higher resistance to frost, fatigue, and deformation [22]. Superpave rutting resistance estimations, based on the $|G^*|/\sin\delta$ ($|G^*|$ is complex shear modulus, δ is phase angle) ratio, are positively affected by the presence of CR. The rubberized bitumen tends to show higher resistance to deformation [23, 24].

When mixed in hot bitumen, CR undergoes a complex interplay of physical and chemical modifications. Billiter et al. [25] reported rubber particle devulcanization, i.e., breaking the sulfur chemical bonding under high-shear conditions. The dimensional changes of CR upon reacting with bitumen were further investigated by Xiao et al. [26]. They noticed that the size reduction of rubber particles was a function of mixing conditions and bitumen type, influencing that process. In a separate study by Venudharan et al. [27], FTIR measurements revealed the formation of bonding between the sulfur moieties contained in the CR and the bitumen.

The properties of CRMB have been reported to be affected by several parameters. Bahia and Davies [28] concluded that the bitumen and CR type influence the high-temperature properties of CRMB. At low temperature, the CRMB is influenced mainly by the bitumen. Furthermore, other authors indicated that the CR production process, particle size distribution, and chemical composition contribute to the CRMB properties [29–31]. Results indicate that, in comparison to cryogenic and mechanical grinding, waterjet technology has a demonstrated propensity to enhance fracture toughness. Aside from the CRMB component, the blending procedure (temperature and time) [32] together with the storage time are also known to influence the properties [33, 34].

According to Khalili et al. [35], the CR's particle size, shape, and content had a substantial impact on the rheological characteristics of the modified bitumen. The flow behavior, elasticity, loading, and temperature dependence of CRMB were assessed by Kim et al. [36]. Results showed that adding CR as a modifier increased bitumen's viscosity, transferred Newtonian flow characteristics into a shear-thinning flow, decreased creep compliance, improved elasticity, increased stiffness and complex modulus at higher temperatures, and at lower temperatures the phase angle was decreased. Mashaan and Karim [37] evaluated how differing CR contents affected the characteristics of the CRMB. The findings demonstrated that the complex shear modulus, storage modulus, and loss modulus were enhanced with the rubber contents. According to Xie et al. [38], employing activated CR enhanced the stiffness modulus and creep characteristics of bitumen at various temperatures.

The use of chemical additives combined with CR has been explored to improve CRMB performance. Trans-polyoctenamer rubber (TOR) has been reported to decrease the CRMB viscosity [39] and to improve CRMB properties [32, 40, 41], including its storage performance [42]. The improvement of CRMB properties by TOR is ascribed to forming of solid cross-linking bonds between CR particles and some fraction of the bitumen [43]. Polyphosphoric acid (PPA) is another popular additive that has been combined with CR powders to improve bitumen behavior [44]. PPA is claimed to induce changes in the bitumen's chemical composition, affecting the saturate, aromatic, resin, and asphaltene (SARA) fraction ratio [45].

Furthermore, synergetic effects between PPA and CR have been reported. Therefore, increasing the percentage of PPA, the study has shown an increase in the material's stiffness. When dynamic mechanical experiments were utilized instead of the conventional empirical techniques, the shift in transition temperature caused by PPA was more visible. Thus, it appears that using a Dynamic Mechanical Analyzer enhances the differences between various bitumens with or without additions by allowing the examination of material characteristics of low temperature modified bitumen under linear settings.

2.4 Barriers in Application of Crumb Rubber Worldwide

Crumb rubber modification of bituminous mixtures or bituminous binder has attained a high level of technology readiness (TRL). Piao et al. [1] have found that the wet process application of crumb rubber on a global scale is at 7–9 TRL grade, indicating that pilot projects have been implemented in the operational environment. It has attained TRL 5–7 using the dry method, which means that the application is partially or entirely industrialized. The use of this waste material is very region specific. As the high TRL values show, the technology is there, however, there are barriers to its use as in some countries the TRL level is quite low. This is a clear indication that there are certain barriers exist for use of this waste material. Poulikakos et al. [2] presented some ideas to lift some of the barriers: (1) involvement of various stakeholders; (2) knowledge regarding handling, preparation and costs as well as the resulting quality; (3) dissemination of knowledge by the scientific community to the practicing professionals; (4) legislation and standards to guide the professionals.

2.5 Future Tendencies

Although the modification of bitumen with end-of-life tires produced crumb rubber appeared more than 50 years ago, it was only in the last two decades that its study has been more intensive, contributing to the improvement of its characteristics and the performance of the bituminous mixtures.

Crumb rubber was initially used in bituminous mixtures by the wet process, later moving to the dry process. In recent years, rubber activation has been applied to enhance its bond with bitumen and thus improve the performance of bituminous mixtures.

This is the expected trend for the crumb rubber, that is, its activation by different processes or by adding different products, thus improving the performance of bituminous mixtures.

3 Objective

In this present study, the objective is to investigate the bitumen modification using CR to improve pavement performance. Four types of CR, obtained from different grinding processes, namely through the mechanical, cryogenic, and waterjet processes, and reacted and activated rubber were used in this work. Three penetration-grade bitumens were used for modification by the CR. Five laboratories performed mechanical and chemical tests on the modified bitumen. The analysis was carried out to realize the effect of different CR technology on CRMB performance.

4 Materials and Methodology

This work studies the effect of four types of CR on modifying three types of bitumens. CR from the mechanical, cryogenic, waterjet grinding processes and RAR were used. CR powders were obtained from a mix of car and truck waste tires with a maximum dimension of about 4–6 mm, depending on the rubber type. Three base bitumen were used, namely 35/50, 50/70, and 70/100 bitumen. Penetration of these bitumens are 46, 55, and 69 mm/10, respectively. The softening point is 53, 51, and 46 °C, respectively.

The characterization of CRMB included the following methods: Scanning Electron Microscopy (SEM), penetration value (EN 1426 [46]), softening point temperature (EN 1427 [47]), viscosity with the CC27 standard spindle and a helical stirrer (Fig. 1), Fourier Transform Infrared Spectroscopy (FTIR) [48], complex shear modulus relies on a Dynamic Shear Rheometer (DSR) [49], and black space diagram.

Two companies supplied the CR, and bitumen modification was done by each participating laboratory. To ensure that all CRMB were produced and tested at the same conditions, a detailed test procedure protocol was produced and applied by all laboratories (identified as Lab1, Lab2, Lab3, Lab4, and Lab5). One of the kernel factors that affect the CRMB performance is the interaction between CR and bitumen, namely the temperature, the digestion time, and the mixing process. The mixing process using low and high shear mixing equipment also affects the performance of the modified bitumen. For any fair comparison, all modified bitumens must be produced through the same process. Thus, CRMBs were produced at 175 °C with a digestion time of 1 h at 3500 rpm in a high shear mixer. One laboratory (Lab5) used a 300-rpm low shear mixer (electric stirrer mixer). The crumb rubber content was 18% for all rubber types. Additionally, 35% for the RAR was used because this RAR content produces CRMB with viscosity similar to 18% of the other CRs. The detailed composition of used CRMB was shown in Table 1.

A direct comparison between CRMB produced with RAR and the CRMB produced with the other CR cannot be made because RAR includes bitumen and filler that affect the CRMB. Nevertheless, these experiments verify how RAR compares to the other CR.

The viscoelastic behavior of CRMB was assessed based on the Dynamic Shear Rheometer (DSR) measurements for calculating the complex shear modulus, $|G^*|$,

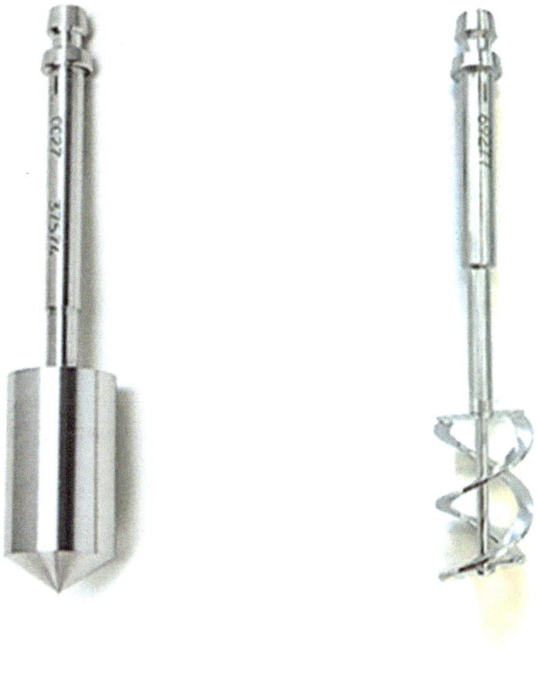

Fig. 1 Standard (CC27) and helical (ST21-2HR-37/120) stirrer (spindle) for viscosity measurements

Table 1 Detailed compositions of the investigated CRMB

Laboratories			1	2	3	4	5
Material							
Binder(s)	Crumb rubber type	%					
70/100	–	0		√	x	√	
	Waterjet (W)	18			x		
	Mechanical (M)	18	√	√	x	√	
	Cryogenic (C)	18	√	√	x	√	
	RAR (R)	18				√	
	RAR (R)	35	√	√		√	
35/50 and 50/70	–	0					√
	Waterjet (W)	18					√
	Mechanical (M)	18					√
	Cryogenic (C)	18					√
	RAR (R)	35					√

*x indicate that only penetration and softening point tests could be performed by this laboratory

and phase angle, δ. Two different geometry plates (8 mm diameter with 2 mm gap, and 25 mm diameter with 1 mm gap) were used to perform temperature and frequency sweep tests for low and high temperatures. Constant strain levels of 0.1% for high temperature range, 34–82 °C, and 0.05% for low-temperature range, 0–40 °C, were used for testing frequencies between 0.1 and 20 Hz.

Viscosity measurements were recorded after homogenizing the CRMB and after it reached a steady state. The test protocol consists of 30 min conditioning time at test temperature followed by 60 min of measurement at a constant speed of 20 rpm with an equivalent shear rate of 26 s^{-1}. At each testing temperature, two samples were examined, and the average of the previous ten readings was noted.

CR was characterized in terms of morphology through SEM images.

The CRMBs are identified with information on base bitumen, CR content, CR type, and laboratory. The CR is identified as M for Mechanical, C for Cryogenic, W for Waterjet, and R for RAR. Thus, a CRMB designated as 70/100 + 18W-Lab2 means that it is made with 70/100 base bitumen, with 18% of Waterjet CR, produced and tested in Laboratory 2.

70/100 base bitumen was used by Lab1, Lab2, Lab3 and Lab4 laboratories, whereas 35/50 and 50/70 base bitumens were only used by laboratory 5 (Lab5).

5 Results

This chapter presents the results obtained in this study, and the resulting analysis deducing the effect of CR technology on the behaviors of CRMB.

5.1 CR Morphology

COXEM M30 plus + scanning electron microscope (SEM) was used to observe the morphology of CRs obtained from different production processes.

The untreated CR images are presented in Fig. 2 for mechanical, cryogenic, waterjet CR at a magnification of 700 times that allows observing all irregularities of the CR surface.

As observed, the morphologies for these three CRs are different. A flat surface characterizes the mechanical grinded CR surface with small granules (around 0.01 mm diameter or less). The mechanical (Fig. 2) and cryogenic (Fig. 2) CR have many small granules observed in the SEM images, while no grains are identified in the waterjet CR. The surface configuration of these CRs will influence the interaction between the bitumen and the CR, and as a result, a more significant interaction with the mechanical and cryogenic CR is expected.

The surface of the treated CR, namely the RAR, can be observed in the SEM images. This CR shows a distinct regularity and smooth surface compared to the untreated CR, characterized by a fluffier texture with many concave and convex units favorable for bitumen adsorption.

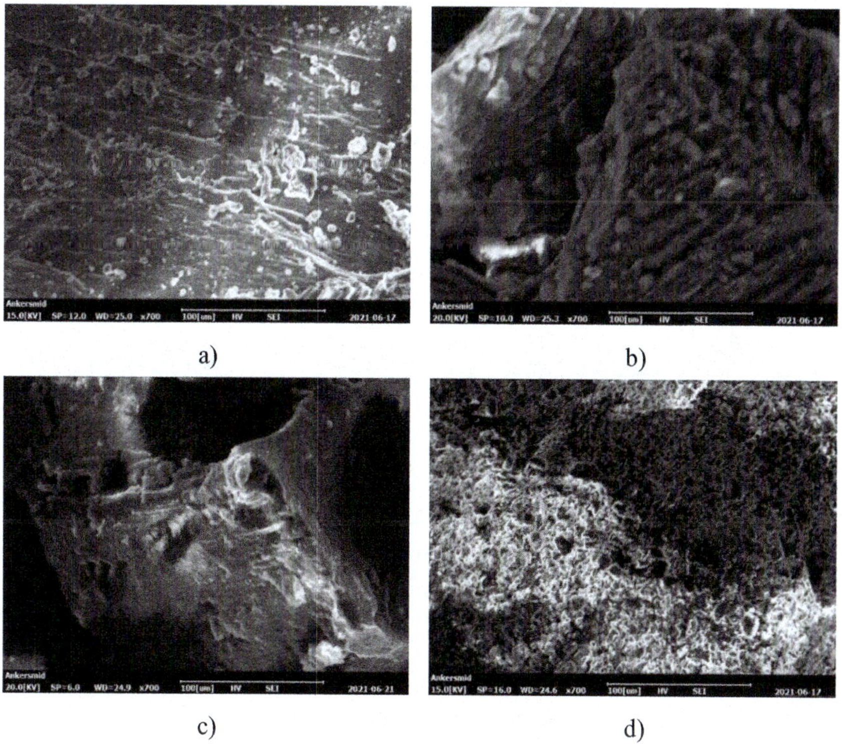

Fig. 2 **a** SEM image of mechanical, **b** cryogenic, **c** waterjet, and **d** RAR. Scale bar = 100 μm

5.2 Penetration

Penetration tests were carried out on CRMB produced with all CR types (mechanical, cryogenic, waterjet, and RAR) and all base bitumens (70/100, 50/70, and 35/50), following the European Standard EN 1426 [46].

Figure 3 shows the results for CRMB produced with 70/100 base bitumen from three labs (Lab2, Lab3 and Lab4). The first observation is related to the base bitumen, where the results from these three labs range from 69 to 108 dmm, showing the reduced reproducibility of this test compared to when tests are done in the same lab.

The CR addition reduces the penetration between 30 and 50%, changing the grade to 35/50. The reduction of the bitumen grade contributes to the increase of the bitumen's stiffness, thus reducing the temperature susceptibility, as shown in [50].

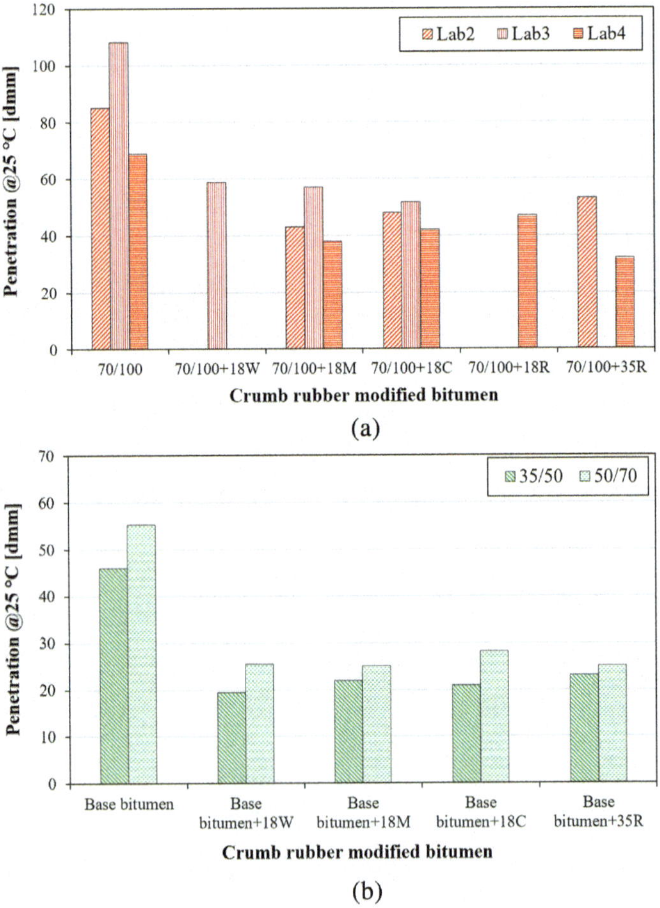

Fig. 3 Penetration of CRMB produced with 70/100 bitumen (**a**) and with 35/50 and 50/70 bitumen by Lab5 (**b**)

The results with other bitumen types, tested by Lab5, shown in Fig. 3, where stiffer bitumen has been used, corroborate this hypothesis. 35/50 and 50/70 bitumen have been modified with 18% of mechanical, cryogenic, and waterjet rubbers and 35% of RAR. As discussed in the previous section, due to additives and filler, the performance of 35% RAR should theoretically be similar to 18% of other types of CR. The stiffening effect is even more prominent as the penetration reduction lies between 45% and 55%. This finding confirms the stiffening effect of all the CRs studied. The penetration reduction also changed the bitumen grade.

The scattering among the same data set for different laboratories is greater than 10%, both for base bitumen and CRMB. For base bitumen, this scatter can be attributed probably to batch differences, while for CRMB the scatter is associated with the production process.

On the other hand, the change in the mechanical properties of the modified bitumen with CR was affected by the swelling of the CR particles in the bitumen, which form a viscous gel that increases the viscosity of the bitumen, as shown in [51]. It can be concluded that the performance depends primarily on the properties of the CR used as reported in [52].

5.3 Softening Point

Conventional test of ring-and-ball softening point was carried out in all laboratories involved in the project following the European Standard EN 1427 [47]. Figure 4 shows the softening point in all CRMB.

The use of CR increases the softening point, implying that all the CRs studied stiffen the base bitumen, corroborating what was found by other researchers [52]. Furthermore, it can be observed that the trend for the softening point is slightly different from that for penetration. For instance, from Fig. 4, it can be observed that the 70/100 bitumen blended with mechanical type CR (70/100 + 18M) and with RAR (70/100 + 35R) showed the highest increase in softening point value compared to the other types of blends. This can indicate that those modifications reduce the tendency of the bitumen to flow at elevated temperatures, as described in [53]. From Fig. 4, it can be observed that the scattering is quite prominent among different laboratories corroborating the hypothesis indicated in the previous section. The softening point increase is higher in the 35/50 bitumen than in the 50/70 bitumen, tested in Lab5. This tendency is less prominent for 50/70 + 35R and for 35/50 + 35R, where the scattering is lower. This can indicate how the CR stiffening potential is even more effective when the CRs are added to a stiffer bitumen. Based on these results, the addition of CR to the bitumen affects the physical properties of all bitumens studied as indicated by the increase in softening point. However, this change is more evident in stiffer bitumens.

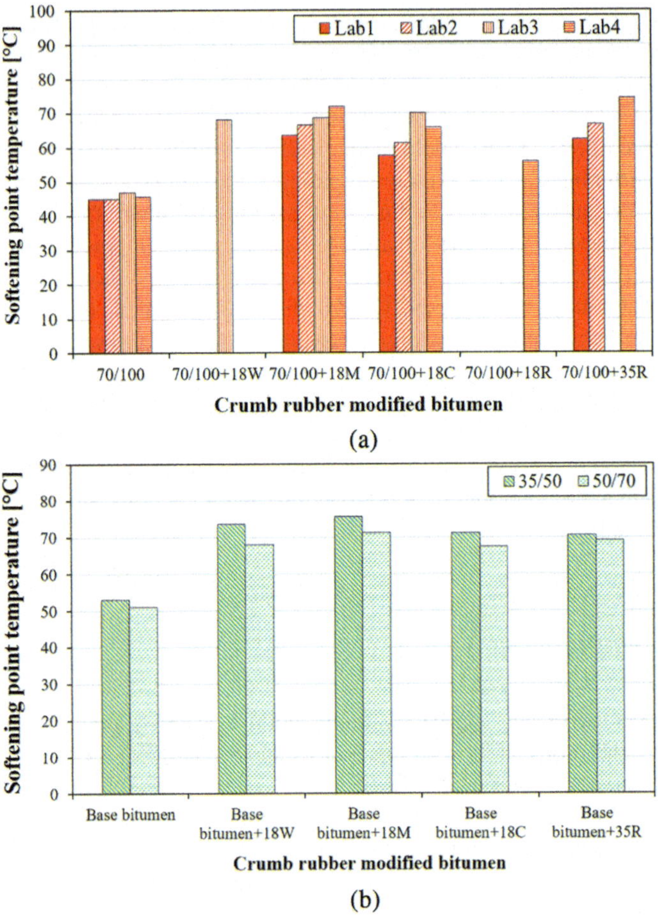

Fig. 4 Softening point of CRMB produced with 70/100 bitumen (**a**) and with 35/50 and 50/70 bitumen by Lab5 (**b**)

5.4 *Viscosity*

Viscosity measurements were carried out using a DSR and a Brookfield viscometer, both in rotational mode. CRMB produced with 70/100 bitumen were tested in the DSR by Lab1, while CRMB made with 50/70 and 35/50 were tested in the Brookfield viscometer by Lab5.

In the DSR, two spindles were used: a helical spindle and a standard spindle, as shown in Fig. 6. The motivation for using two spindles was the assumption that a helical spindle can facilitate the distribution of particles in the bitumen and results in a more homogeneous fluid and more accurate viscosity measurements. For comparative analysis, only 70/100 + 18M and 70/100 + 18C were tested with the helical spindle. However, using the standard spindle (CC27), 70/100 + 35R was tested, in addition to 70/100 + 18M and 70/100 + 18C.

Figure 5 presents the viscosity results of 70/100 + 35R, 70/100 + 18M and 70/100 + 18C CRMB using the standard spindle (CC27). The results show that 70/100 + 35R has the highest viscosity followed by 70/100 + 18M and 70/100 + 18C, in agreement with the crumb rubber texture observed in SEM images. A proportionate drop in viscosity for all modified bitumens (i.e., 70/100 + 35R, 70/100 + 18M and 70/100 + 18C) was observed with the increase in test temperature from 135 °C to 170 °C.

Figure 6 presents the viscosity results of 70/100 + 18M and 70/100 + 18C using helical and standard spindles. The terms SS and HS refer to the standard spindle and helical spindle, respectively. The results reveal a higher viscosity value of both 70/100 + 18M and 70/100 + 18C CRMB at test temperature (135 °C, 150 °C, 160 °C and 170 °C) using helical spindle compared with standard spindle. The higher viscosity values of 70/100 + 18M and 70/100 + 18C (see Fig. 6) using helical spindle could

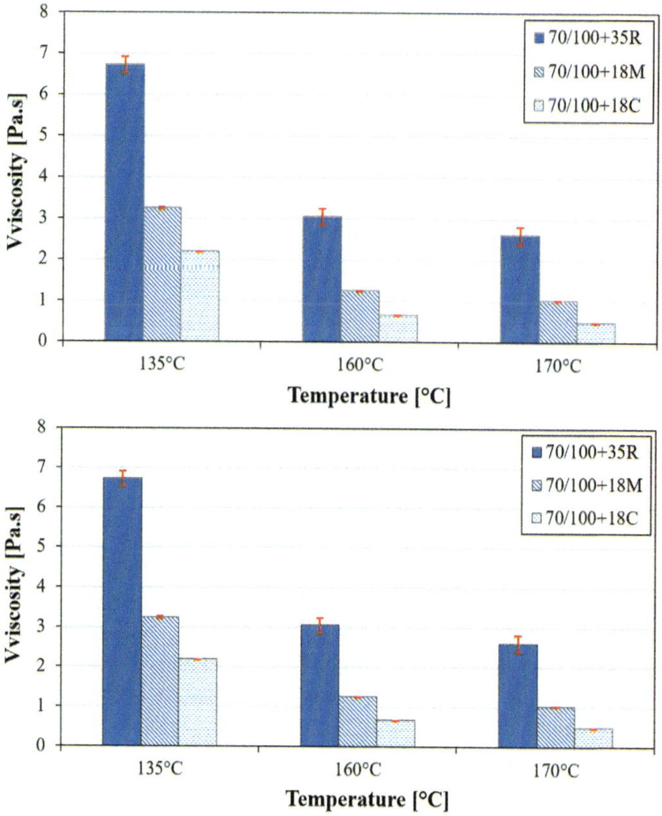

Fig. 5 Viscosity results of 70/100 CRMB using the standard spindle (CC27) by Lab1

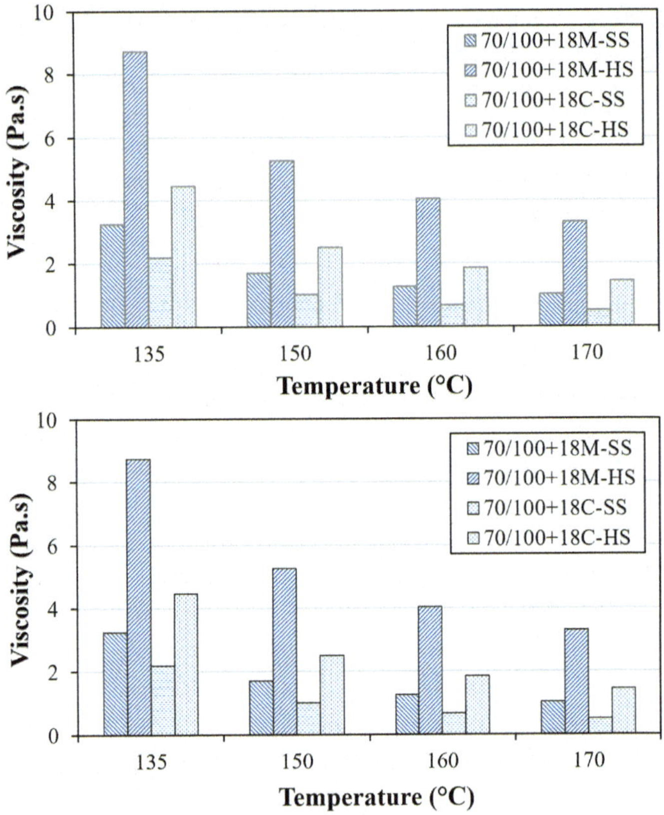

Fig. 6 Viscosity results of 70/100 CRMBs (standard vs. helical spindle) by Lab1

be attributed to the fact that a large contact area of the helical spindle resulted in higher values of viscosity compared to the results of the standard spindle.

Furthermore, the higher viscosity with the helical spindle might also be because CRMB is better homogenized than with the standard spindle. The viscosity with the standard spindle results from the interaction of the spindle and a thin narrow bitumen wall around the standard spindle, and there is a chance that CR particles are concentrated at the bottom part of the container. However, for the Helical spindle, CR particles might get the opportunity of being homogenized due to the spiral nature of the spindle during the viscosity measurements. Overall, based on the results obtained from viscosity measurements, using the helical spindle for CRMB is a logical option, and more research and development in this area is encouraged.

The results for 35/50 and 50/70 CRMB are presented in Fig. 7 for temperatures ranging from 100 °C to 200 °C for cryogenic, mechanical and waterjet CR. The typical evolution of the viscosity with temperature is visible. Minor irregularities in

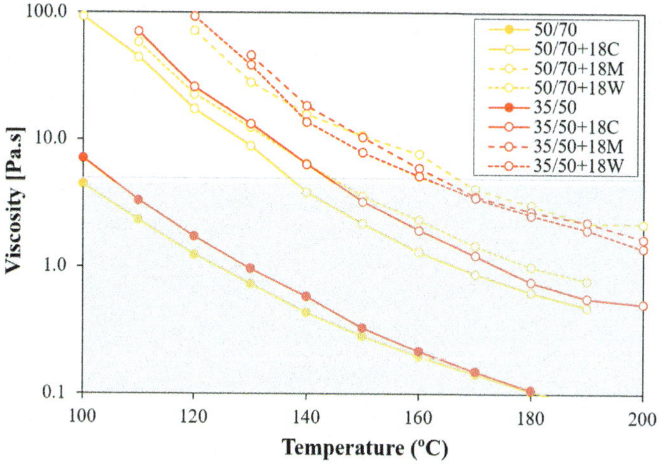

Fig. 7 Viscosity results for 35/70 and 50/70 CRMBs by Lab5

the viscosity evolution are due to the CRMB temperature stabilization and homogeneity. However, it is clear that there is a reduction of the CRMB viscosity with increasing temperature.

Furthermore, the effect of the CR type on the bitumen is also visible. Due to its surface, mechanical CR has the most significant interaction with bitumen, producing the highest viscosity CRMB. On the contrary, cryogenic CR has reduced interaction with the bitumen. Despite its surface type, waterjet CRMB has a viscosity between mechanical and cryogenic CRMB.

Following ASTM D6114 [54] standard for CRMB, where the 5 Pa·s is defined as the viscosity limit, these CRs can be used as bitumen modifiers for temperatures starting at 140 °C.

5.5 *FTIR*

Fourier Transform Infrared Spectroscopy (FTIR) methodology was used for assessing the changes in bitumen due to its modification by CR additives through the quantification of the chemical interaction of certain elements. The principle of this analysis is based on the absorption of the electromagnetic energy by the chemical bonds for a wide range of wavelengths.

In this work, FTIR analysis was combined with the Attenuated Total Reflection (ATR). The analysis was carried out at 20 °C. Wavelengths between 4000 and 550 cm^{-1} were tested with a resolution of 4 cm^{-1}. The obtained spectra were analyzed to verify how the chemical bonds change due to the presence of CR.

FTIR was carried out on 70/100 CRMB, tested by Lab4, and results are shown in Fig. 8. Each spectrum is the average of three replications. Base bitumen and CRMBs

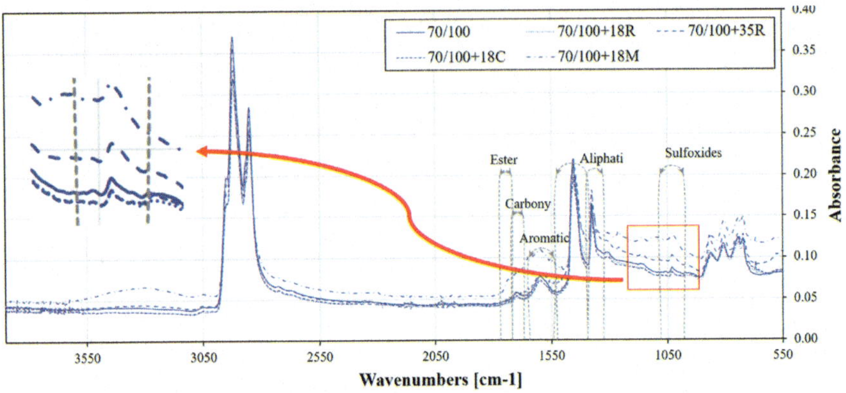

Fig. 8 FTIR spectra of CRMBs results from Lab4

present similar spectrum, mainly in the bands used for the characterization of the bitumen, except for cryogenic CRMB in the sulfoxide area.

The analysis of FTIR spectra was carried out by analyzing the following bands:

1. Around the 1700 cm^{-1} where it is defined the carbonyl area.
2. Around 1030 cm^{-1} where it is defined the sulfoxide area.
3. Below 3000 cm^{-1} (stretching area) and 1500 cm^{-1} (bending area), where the CH2 and CH3 groups are included.
4. Around 1460 cm^{-1}, so called symmetric aliphatic group.
5. Around 1376 cm^{-1}, so called asymmetric bending vibrations group.

Comparing the carbonyl and sulfoxide groups, the ageing of all samples is small. The degree of oxidation of CRMB can be evaluated quantitatively by the development of C=O (carbonyl) and S=O (sulfoxide) groups. Bands around 1700 cm^{-1} proved to be a stretching vibration of C=O, and bands around 1000 cm^{-1} affect S=O vibration. From the FTIR diagram, three areas are calculated using fixed wavelength numbers: (i) from 1670 to 1720 cm^{-1} for carbonyl area; (ii) from 1325 to 1525 cm^{-1} for the aliphatic area; (iii) from 980 to 1080 cm^{-1} for sulfoxides area. The indexes $I_{c=o}$ and $I_{s=o}$ were calculated by dividing the corresponding carbonyl and sulfoxides areas by the stretching vibrations of the aliphatic areas. Figure 9 shows the FTIR indexes, namely the FTIR carbonyl (ICO) and FTIR sulfoxide (ISO).

Usually, indexes are used to characterize ageing or chemical modification. As it can be seen from Fig. 9, no excessive ageing of the bitumen, except for cryogenic CRMB. However, it can be seen that 70/100 + 35R has a higher $I_{s=o}$, which could be related to the presence of filler. In particular, it can be distinguished among CRMB 70/100 + 18R and 70/100 + 35R. The values for indexes are within the precision of the equipment, and the only significant difference is for 70/100 + 18C. This FTIR analysis means that the interaction between bitumen and CR is mainly a physical phenomenon.

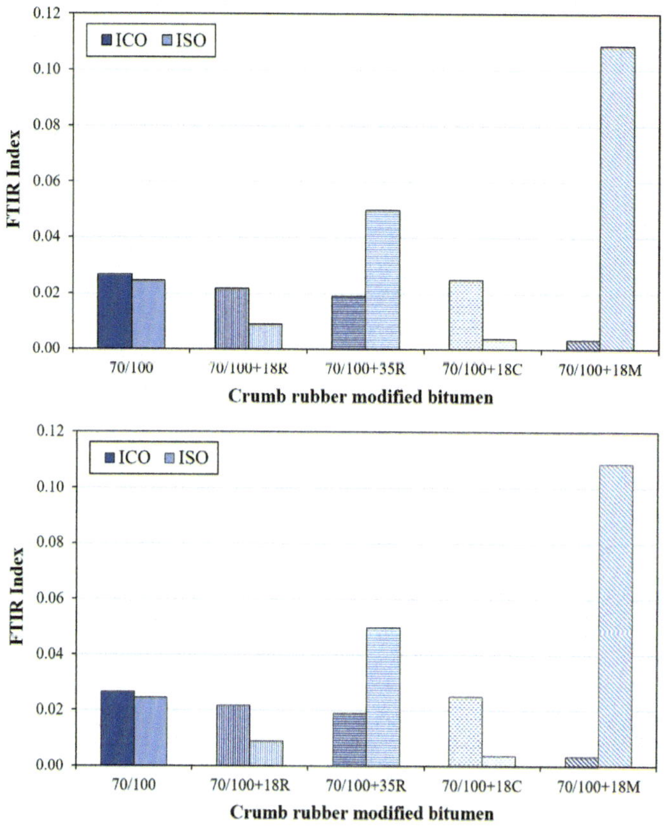

Fig. 9 Carbonyl and sulfoxide indexes for FTIR spectra

5.6 *Shear Complex Modulus*

In this section, the outcome of the shear complex modulus tests is reported. The results are split into two blocks because of the difference in bitumens and CR materials used by the participating laboratories. The first one is related to the use of 70/100 bitumen (laboratories: Lab1, Lab2, and Lab4), and the other is associated with the 35/50 and 50/70 bitumens (laboratory: Lab5); this approach returns a more consistent analysis.

The first inspection of the shear modulus data is performed on the G* master curves as these provide a broader visualization of the material response across temperature and frequency ranges. Master curves were generated according to the sigmoidal model equation. Reference shifting temperature was set to 22 °C, and the shift factor was calculated based on the well-known Williams Landel and Ferry equation [55]. Figures 10, 11 and 12 present the |G*| master curves.

The first observation of the $|G^*|$ master curves allows identifying the clockwise rotation of the master curve. CR's incorporation provides a stiffening effect

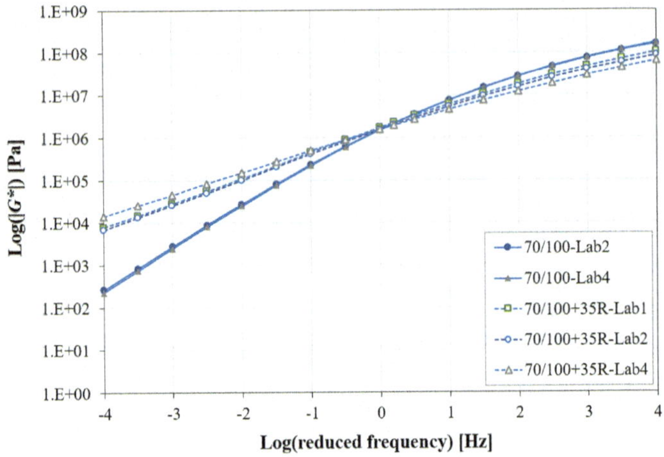

(a)

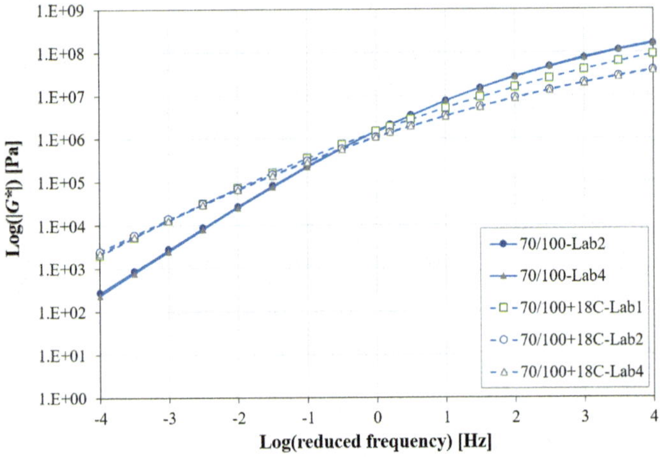

(b)

Fig. 10 $|G^*|$ master curves for CRMB produced with 70/100 base bitumen: **a** with 35% of RAR, and **b** with 18% of cryogenic CR

of the CRMB in the high-temperature/low-frequency domain. Simultaneously, a comparable decrease in stiffness is observed on the other end of the spectrum (low-temperature/high-frequency). These effects are verified for all CRMB.

When considering the base bitumen 70/100, bitumens with 35% RAR and, to a minor extent, mechanical CRMB appear remarkably affect $|G^*|$ with an increase of about half order of magnitude at the lowest frequency and highest temperature in a range of one order of magnitude of frequency. The overall trend among the different CR materials is confirmed by Lab4 results, where an additional bitumen incorporating RAR at 18% is investigated. The consistency of the measurements

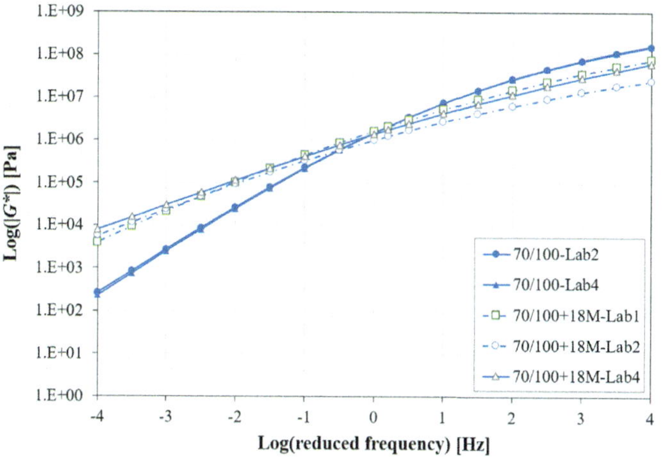

(a)

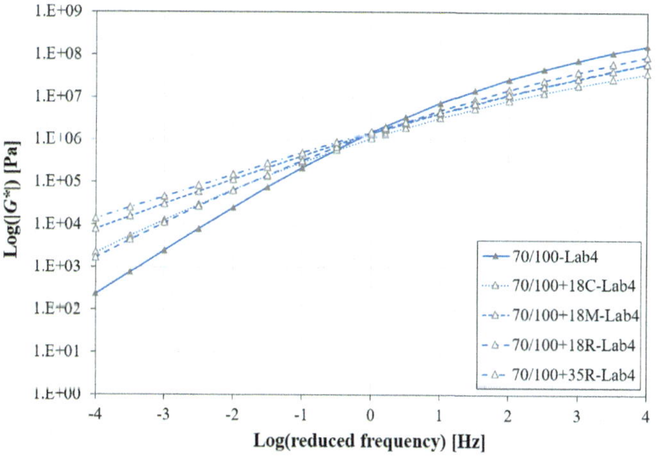

(b)

Fig. 11 |*G**| master curves for CRMB produced with 70/100 base bitumen: **a** with 18% of mechanical CR, and **b** all CRMBs produced with 70/100 base bitumen by Lab4

across the three laboratories (Lab1, Lab2, and Lab4) using the same core 70/100 bitumen appears reasonable.

Concerning the results from Lab5, two different bitumens having penetration grading 35/50 and 50/70 were used. So significant differences are observed between the two bitumens when CRs are mixed with the original bitumen, except the |*G**| reduction for high-frequency/low-temperature is more remarkable for the 50/70 CRMB.

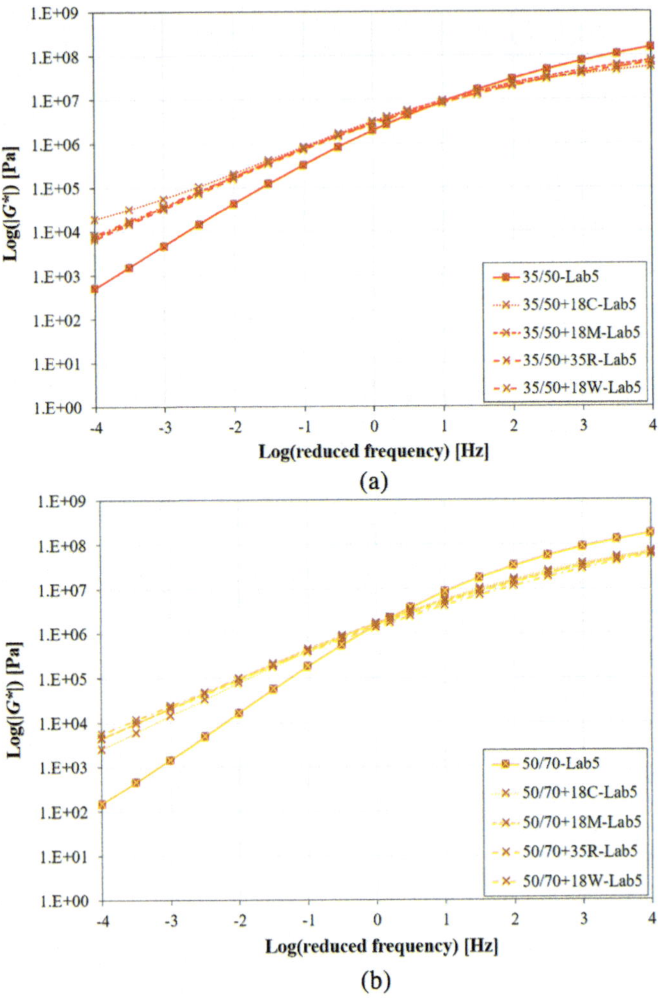

Fig. 12 $|G^*|$ master curves for all CRMBs produced: **a** with 35/50 base bitumen, and **b** with 50/70 base bitumen

For the harder bitumen (35/50), the CR appears to substantially increase the overall stiffness in the low-frequency/high-temperature range. The effect of crumb rubber type is not visible except for 35/50 bitumen modified by RAR.

Given the master curves results, additional analysis was performed by restricting the attention to specific values of the shear complex modulus. For this purpose, the shifting temperatures of $T = 22$ and 58 °C and a conventional frequency of 1.59 Hz were set as reference; the shear complex modulus $|G^*|$ at 1.59 Hz, together with Superpave Performance Grade (PG) parameters for rutting ($|G^*|/\sin\delta$) at $T = 58$ °C and fatigue (|G^*|sinδ) at $T = 22$ °C were evaluated for the different CRs. Here the ageing procedure commonly performed before the tests was not performed.

A visual synthesis of the results is presented in Figs. 13, 14, and 15 for 70/100, 35/50, and 50/70 bitumens.

Considering the three laboratories using the same base 70/100 bitumen, also at $T = 22$ °C, good data consistency is observed for Lab2 and the Lab4, while overall higher (for |G^*| and |G^*|/sinδ) or lower (|G^*|sinδ) values are exhibited from the measurements performed at Lab1. As can be observed in the master curves, the values of |G^*| and |G^*|/sinδ for the pure 70/100 bitumen is overall moderately higher than the bitumen incorporating any of the CRs, while an opposite trend is experienced for the rutting parameter (|G^*|/sinδ). RAR at 35% content seems to provide the best material response in fatigue and rutting among the different CR additives. This is not surprising as a substantial elastic contribution is provided to the bituminous material

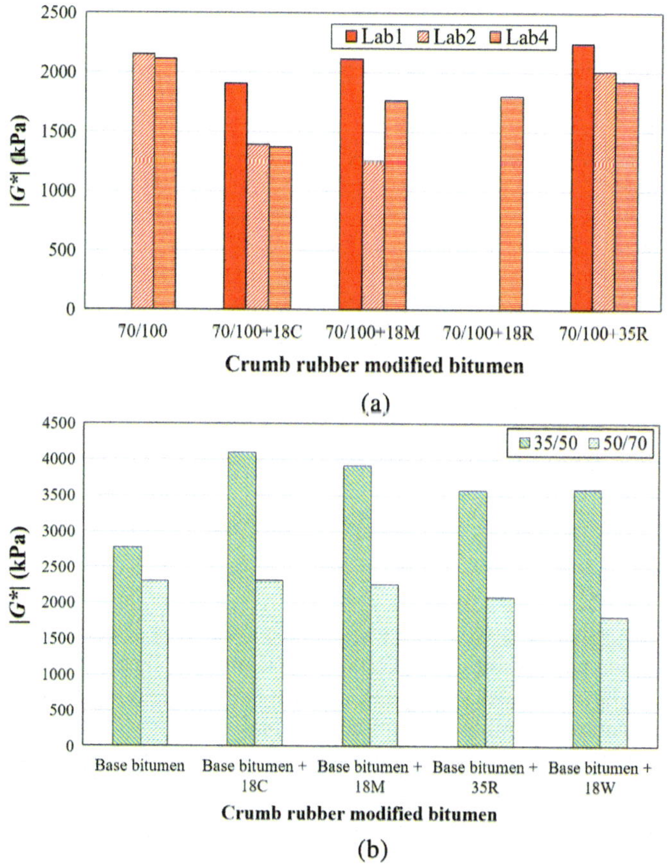

Fig. 13 |G^*| at $T = 22$ °C and $f = 1.59$ Hz: **a** 70/100 bitumen group; **b** 35/50 and 50/70 bitumen group

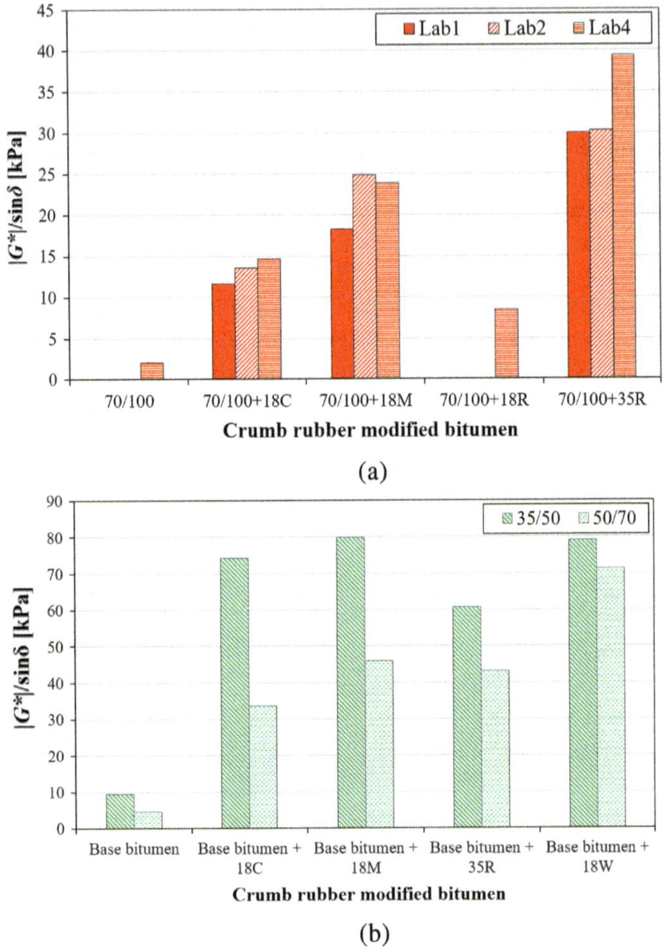

Fig. 14 $|G^*|/\sin\delta$ at $T = 58$ °C and $f = 1.59$ Hz: **a** 70/100 bitumen group; **b** 35/50 and 50/70 bitumen group

by the RAR, as this CR also contains filler and additives, as discussed earlier. When considering the DSR results from Lab5, a discriminating factor is provided by the difference in bitumen used. In this case, the CR-bitumen composites prepared with 35/50 bitumen are overall returning higher values for $|G^*|$, $|G^*|/\sin\delta$ and $|G^*|\sin\delta$. This was not necessarily obvious from the master curves. It seems to suggest that the bitumen's penetration grade ultimately dictates the material response, at least at the specifically selected temperatures.

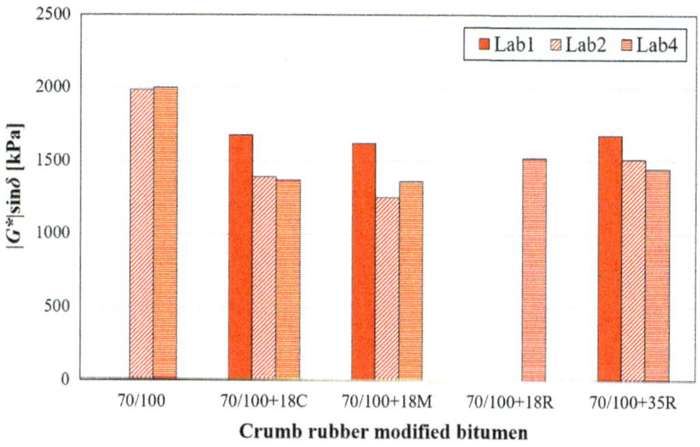

(a)

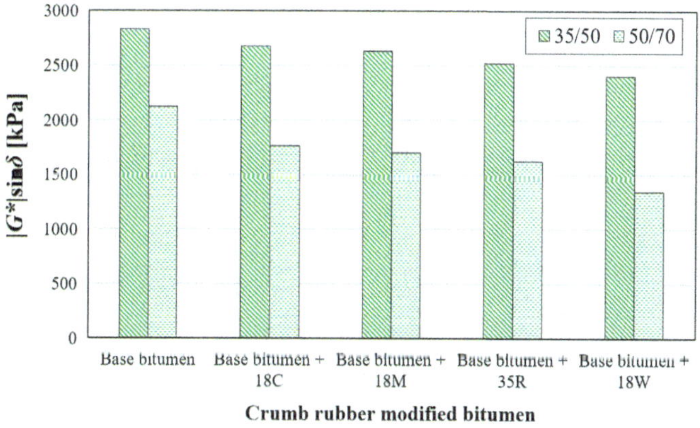

(b)

Fig. 15 $|G^*|\sin\delta$ at $T = 22$ °C and $f = 1.59$ Hz: **a** 70/100 bitumen group; **b** 35/50 and 50/70 bitumen group

5.7 Black Space Diagram

Bitumen mechanical behavior is usually expressed by the complex shear modulus, $|G^*|$, and its corresponding phase angle, δ. These values are utilized to construct the master curve of $|G^*|$ that allows the characterization of the modulus over a wide range of temperatures and frequencies. Different modes are used for presenting the phase angle, the black diagram curve, $|G^*|$ versus δ, being the most known method. In the black diagram, base bitumen is characterized by having a shape of a quarter of an ellipse with a fixed point for zero shear modulus and 90° phase angle.

Polymer modified bitumen is characterized by having a black diagram with the shape of half of an ellipse [56]. CRMB presents a black diagram with two ellipse segments [57, 58], as observed in Fig. 16 for cryogenic CRMB (legend represents

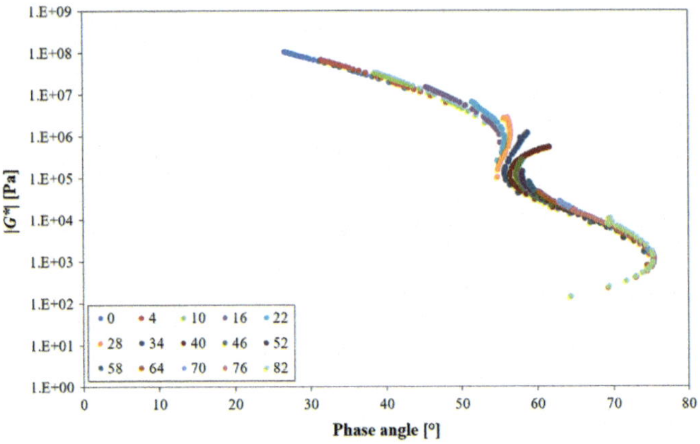

Fig. 16 Typical black diagram for CRMB with numbers and colors indicating test temperature

testing temperatures). This shape of the black diagram can be expressed by Eq. (1), where X is the logarithm of the complex shear modulus ($X = \log(|G^*|)$) and a, b, c, d, e and f are statistically determined coefficients. This equation is only valid for the range of values of the black diagram.

$$phase\ angle = \frac{1}{a + bX + cX^2 + dX^3 + eX^4 + fX^5} \tag{1}$$

The modelling of all black diagram curves through Eq. (1) is represented in Fig. 17. It is visible that a maximum phase angle peak characterizes the curve at low complex shear modulus, a constant phase angle at the middle of the curve, and an increase of phase angle for high complex shear modulus.

The black diagrams in Fig. 17 show that all CRMBs displayed a characteristic constant phase angle at lower complex modulus, demonstrating that they grow less dependent on time. As temperature rises, the complex modulus' viscous component takes the dominant stage. The time–temperature superposition principle (TTSP) does not hold true for this material at higher temperatures in this situation since the influence of CR can be observed in the phase angle as the curves diverge from the distinctive conventional trend found in bitumen. When using TTSP on viscoelastic materials, the material should be seen as thermo-rheological simple, meaning that temperature simply affects the molecules' Brownian motion. This suggests that the structure is unaffected by temperature. While in CRMB, greater temperatures preclude the assumption of this thermo-rheological simple behavior. The findings of the phase angle show this divergence. The solid CR particles regulate the rheology when temperatures are greater. These findings support those made public in the publication [59].

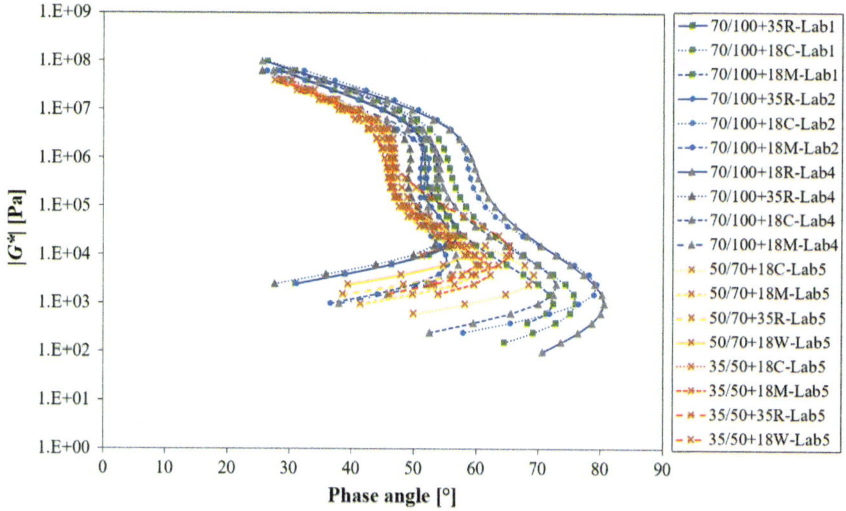

Fig. 17 Black diagrams obtained with the testing program

Black diagrams show a considerable dispersion for all tested CRMBs. This variation is significant for low complex shear modulus, where the phase angle has variations of more than 50°. The phase angle varies around 10° for high complex shear modulus, reducing as the complex shear modulus increases. The constant phase angle appears for the complex shear modulus between 10^5 and 10^6 Pa and the inflexion point between 10^6 and 10^7 Pa. Furthermore, from Fig. 17, it can be deduced that as expected, the temperature directly affects the behaviour of the CRMB. Both materials, bitumen and CR, are stiff and elastic in the cold temperature (high frequency) domain. In contrast, the bitumen becomes soft and viscoelastic in the high temperature (low frequency) domain, whereas the rubber remains elastic. In this domain, more dispersion in the results is seen as the base bitumen properties and the rubber properties diverge.

The black diagrams for cryogenic CRMBs of this work are represented in Fig. 18, where there is a clear separation between 70/100 CRMBs and the other two. It seems that the maximum phase angle in the black diagram is a function of the grade of the base bitumen. As the penetration increases, the maximum phase angle moves to the right (toward lower values of the complex modulus).

The black diagram curves for 35/50 and 50/70 bitumen are almost coincidental because while 35/50 bitumen has a penetration of 46 mm/10, the 50/70 has a penetration of 56 mm/10. However, there is a difference in the cryogenic CR's maximum phase angle location, which does not appear for the other CRs. The different results show some differences for low complex shear modulus where the difference can reach 10° in the phase angle.

Figure 19 presents the effect of the CR type on the black diagram obtained for 35/50 base bitumen. The CR's primary impact is on the behaviour at low complex shear

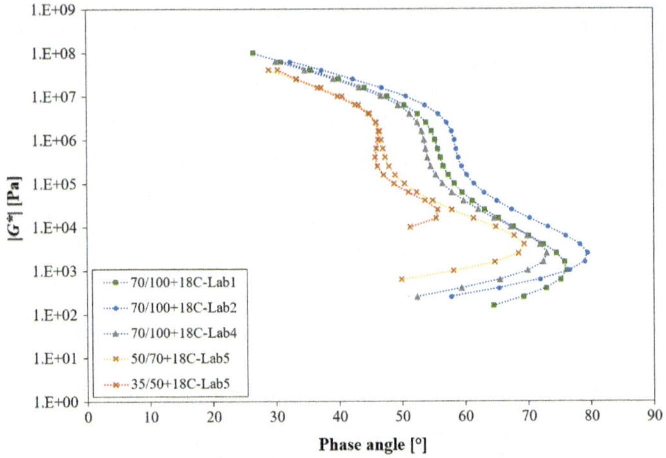

Fig. 18 Influence of the bitumen type on black diagram

modulus, where the curve peak is a function of the CR type. This peak increases as CR surface presents more irregularities. For high complex shear modulus, the behaviour is similar for all CRs.

The RAR produced a different behaviour in the black diagram because, on the one hand, it contains filler and additives, and on the other hand, the CRMB was produced with only one minute of interaction with the bitumen. However, in the high shear modulus regime, it has a similar behaviour as the other CRs.

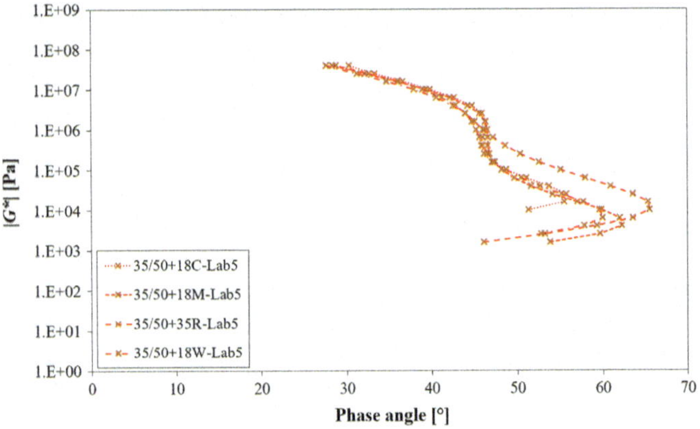

Fig. 19 Influence of CR type on black diagram

6 Summary and Conclusions

Interlaboratory experiments were performed on four types of CR and three types of bitumen, and their effect on the mechanical and chemical properties of CRMB was investigated. The results showed that:

- FTIR analysis concluded that the interaction between bitumen and CR is mainly a physical phenomenon.
- There was a general agreement between the master curves from the different laboratories. The incorporation of CR material seems to provide a stiffening effect to the CRMB in the high-temperature/low-frequency domain. Simultaneously, a comparable decrease in stiffness is observed on the other end of the spectrum (low-temperature/high-frequency).
- Despite some differences in the non-mechanical tests, i.e., penetration, softening point and viscosity, the results of the mechanical tests (complex shear modulus) suggest that the bitumen's penetration grade ultimately dictates the CRMB response.

7 Perspectives and Outlook

The testing programme in this work provided essential conclusions about the bitumen modification by crumb rubber from end-of-life tires.

These conclusions provide a pathway for more research on the CRMB's physical and mechanical properties. But it is in the chemical field that the investigation should focus on understanding the interaction between the rubber and the bitumen. FTIR has been used to study this effect, but results are minimal, and no essential conclusions, mainly practical conclusions, have been obtained.

Another focus for research is the emerging treated crumb rubber for the dry process. The mechanism for bitumen modification is different from the wet process, and it is vital to understand how it works. For this type of crumb rubber, research should be carried out not only on the chemical effects of the modification but on how rubber modification affects the physical and mechanical properties.

An additional aspect to consider for further analysis is the link across the different testing methods adopted during the investigation: conventional, rheological, chemical, and microscopy. Generating trends and connections from results obtained from various tools and disciplines might be challenging with a traditional statistics approach. Artificial Intelligence and Machine Learning have a high potential for revealing hidden relationships, enabling more sophisticated links across materials' properties. In this sense, an ongoing analysis has been started using Artificial Neural Network (ANN) solutions.

Therefore, despite the solid knowledge developed during the last twenty years of studying CRMB, some aspects of the rubber-bitumen interaction need to be deeply studied from different perspectives.

Acknowledgements The authors would like to thank Anton Paar for providing the helical spindle for these investigations.

References

1. Piao Z, Mikhailenko P, Kakar MR, Bueno M, Hellweg S, Poulikakos LD (2021) Urban mining for asphalt pavements: a review. J Clean Prod 280:124916. https://doi.org/10.1016/j.jclepro.2020.124916
2. Poulikakos LD, Papadaskalopoulou C, Hofko B, Gschösser F, Cannone Falchetto A, Bueno M, Arraigada M, Sousa J, Ruiz R, Petit C et al (2017) Harvesting the unexplored potential of European waste materials for road construction. Resour Conserv Recycl 116:32–44. https://doi.org/10.1016/j.resconrec.2016.09.008
3. Cann G NZ's tyre mountains keep growing in the absence of recycling scheme. https://www.stuff.co.nz/environment/92515136/millions-of-tyres-becoming-unrecyclable-in-absence-of-recycling-scheme. Accessed 11 July 2022
4. Tasalloti A, Chiaro G, Murali A, Banasiak L, Palermo A, Granello G (2021) Recycling of end-of-life tires (ELTs) for sustainable geotechnical applications: a New Zealand perspective. Appl Sci 11:7824. https://doi.org/10.3390/app11177824
5. Afrin H, Huda N, Abbasi R (2021) Study on end-of-life tires (ELTs) recycling strategy and applications. In: IOP conference series: materials science and engineering, vol 1200, p 012009. https://doi.org/10.1088/1757-899x/1200/1/012009
6. Bianco I, Panepinto D, Zanetti M (2021) End-of-life tyres: comparative life cycle assessment of treatment scenarios. Appl Sci 11:3599. https://doi.org/10.3390/app11083599
7. Roberts F, Lytton R (1987) FAA mixture design procedure for asphalt-rubber concrete. Transp Res Rec 1115:216–225
8. Abang Ismawi Hassim DH, Abraham F, Summerscales J, Brown P (2019) The effect of interface morphology in waste tyre rubber powder filled elastomeric matrices on the tear and abrasion resistance. Express Polym Lett 13:248–260. https://doi.org/10.3144/expresspolymlett.2019.21
9. Abang Ismawi Hassim DH, Abraham F, Summerscales J, Brown P (2018) Fatigue properties and fracture morphology of micronised rubber powder (MRP) from waste tyres in unfilled elastomers. Int J Mater Mech Manuf 6:281–285. https://doi.org/10.18178/ijmmm.2018.6.4.392
10. Shen J, Amirkhanian S, Xiao F, Tang B (2009) Influence of surface area and size of crumb rubber on high temperature properties of crumb rubber modified binders. Constr Build Mater 23:304–310. https://doi.org/10.1016/j.conbuildmat.2007.12.005
11. Kedarisetty S, Biligiri KP, Sousa JB (2016) Advanced rheological characterization of reacted and activated rubber (RAR) modified asphalt binders. Constr Build Mater 122:12–22. https://doi.org/10.1016/j.conbuildmat.2016.06.043
12. Sousa JB, Vorobiev A, Rowe GM, Ishai I (2013) Reacted and activated rubber: elastomeric asphalt extender. Transp Res Rec 2371:32–40. https://doi.org/10.3141/2371-04
13. Rath P, Meister J, Jahangiri B (2020) Evaluation of engineered crumb rubber (ECR) performance characteristics, including warm-mix equivalence with polymer, draindown prevention, and release enhancement. In: RILEM international symposium on bituminous materials (ISBM Lyon 2020). Springer, Cham
14. Mashaan NS, Ali AH, Karim MR, Abdelaziz M (2014) A review on using crumb rubber in reinforcement of asphalt pavement. Sci World J 2014:214612. https://doi.org/10.1155/2014/214612
15. Tefera BY, Tadele K, Geremew A (2018) Evaluation of the effect of rubber modified bitumen on asphalt performance. Am J Civil Eng 6:88–93. https://doi.org/10.11648/j.ajce.20180603.11
16. Nejad FM, Aghajani P, Modarres A, Firoozifar H (2012) Investigating the properties of crumb rubber modified bitumen using classic and SHRP testing methods. Constr Build Mater 26:481–489. https://doi.org/10.1016/j.conbuildmat.2011.06.048

17. Navarro FJ, Partal P, Martínez-Boza F, Gallegos C (2004) Thermo-rheological behaviour and storage stability of ground tire rubber-modified bitumens. Fuel 83:2041–2049. https://doi.org/10.1016/j.fuel.2004.04.003
18. Dantas Neto SA, Farias MM, Pais JC, Pereira PAA, Sousa JB (2006) Influence of crumb rubber and digestion time on the asphalt rubber binders. Road Mater Pavement Des 7:131–148. https://doi.org/10.1080/14680629.2006.9690030
19. Cong P, Xun P, Xing M, Chen S (2013) Investigation of asphalt binder containing various crumb rubbers and asphalts. Constr Build Mater 40:632–641. https://doi.org/10.1016/j.conbuildmat.2012.11.063
20. Hınıslıoğlu S, Ağar E (2004) Use of waste high density polyethylene as bitumen modifier in asphalt concrete mix. Mater Lett 58:267–271. https://doi.org/10.1016/S0167-577X(03)00458-0
21. Dantas-Neto SA, Farias MM, Pais JC, Pereira PAA (2006) Dense graded hot mixes using asphalt-rubber binders with high rubber contents. Road Mater Pavement Des 7:29–46. https://doi.org/10.1080/14680629.2006.9690025
22. Oikonomou N, Mavridou S (2009) The use of waste tyre rubber in civil engineering works. In: Khatib JM (ed) Sustainability of construction materials. Woodhead Publishing. https://doi.org/10.1533/9781845695842.213
23. Sutanto M, Bala N, Al Zaro K, Sunarjono S (2018) Properties of crumb rubber and latex modified asphalt binders using superpave tests. In: MATEC web of conferences, vol 203, p 05007. EDP Sciences. https://doi.org/10.1051/matecconf/201820305007
24. Khatijah Abu Bakar S, Ezree Abdullah M, Mustafa Kamal M, Abd Rahman R, Buhari R, Putra Jaya R, Sabri S, Abdullahi Ahmad K (2018) The effect of crumb rubber on the physical and rheological properties of modified binder. J Phys Conf 1049:012099. https://doi.org/10.1088/1742-6596/1049/1/012099
25. Billiter TC, Davison RR, Glover CJ, Bullin JA (1997) Production of asphalt-rubber binders by high-cure conditions. Transp Res Rec 1586:50–56. https://doi.org/10.3141/1586-07
26. Xiao F, Putman BJ, Amirkhanian SN (2006) Laboratory investigation of dimensional changes of crumb rubber reacting with asphalt binder. In: Proceedings of the asphalt rubber pavement conference, Cathedral City, California, USA, 25–27 Oct 2006
27. Venudharan V, Biligiri KP, Das NC (2018) Investigations on behavioral characteristics of asphalt binder with crumb rubber modification: rheological and thermo-chemical approach. Constr Build Mater 181:455–464. https://doi.org/10.1016/j.conbuildmat.2018.06.087
28. Bahia HU, Davis R (1994) Effect of crumb rubber modifiers (CRM) on performance related properties of asphalt binders. J Assoc Asphalt Paving Technol 63:414–449
29. Liu G, Liang Y, Chen H, Wang H, Komacka J, Gu X (2019) Influence of the chemical composition and the morphology of crumb rubbers on the rheological and self-healing properties of bitumen. Constr Build Mater 210:555–563. https://doi.org/10.1016/j.conbuildmat.2019.03.205
30. Loderer C, Partl MN, Poulikakos LD (2018) Effect of crumb rubber production technology on performance of modified bitumen. Constr Build Mater 191:1159–1171. https://doi.org/10.1016/j.conbuildmat.2018.10.046
31. Plemons CD (2013) Evaluation of the effect of crumb rubber properties on the performance of asphalt binder. Master Thesis, Auburn University, Alabama, USA
32. Li B, Huang W, Tang N, Hu J, Lin P, Guan W, Xiao F, Shan Z (2017) Evolution of components distribution and its effect on low temperature properties of terminal blend rubberized asphalt binder. Constr Build Mater 136:598–608. https://doi.org/10.1016/j.conbuildmat.2017.01.118
33. Pais J, Lo Presti D, Santos C, Thives L, Pereira P (2019) The effect of prolonged storage time on asphalt rubber binder properties. Constr Build Mater 210:242–255. https://doi.org/10.1016/j.conbuildmat.2019.03.155
34. Thives LP, Pais JC, Pereira PAA, Trichês G, Amorim SR (2013) Assessment of the digestion time of asphalt rubber binder based on microscopy analysis. Constr Build Mater 47:431–440. https://doi.org/10.1016/j.conbuildmat.2013.05.087
35. Khalili M, Jadidi K, Karakouzian M, Amirkhanian S (2019) Rheological properties of modified crumb rubber asphalt binder and selecting the best modified binder using AHP method. Case Stud Constr Mater 11:e00276. https://doi.org/10.1016/j.cscm.2019.e00276

36. Kim HS, Lee SJ, Amirkhanian S (2010) Rheology investigation of crumb rubber modified asphalt binders. KSCE J Civ Eng 14:839–843. https://doi.org/10.1007/s12205-010-1020-9
37. Mashaan NS, Karim MR (2013) Investigating the rheological properties of crumb rubber modified bitumen and its correlation with temperature susceptibility. Mater Res 16:116–127. https://doi.org/10.1590/S1516-14392012005000166
38. Xie J, Yang Y, Lv S, Zhang Y, Zhu X, Zheng C (2019) Investigation on rheological properties and storage stability of modified asphalt based on the grafting activation of crumb rubber. Polymers 11:1563. https://doi.org/10.3390/polym11101563
39. Liu H, Chen Z, Wang W, Wang H, Hao P (2014) Investigation of the rheological modification mechanism of crumb rubber modified asphalt (CRMA) containing TOR additive. Constr Build Mater 67:225–233. https://doi.org/10.1016/j.conbuildmat.2013.11.031
40. Ng Puga KLN, Williams RC (2016) Low temperature performance of laboratory produced asphalt rubber (AR) mixes containing polyoctenamer. Constr Build Mater 112:1046–1053. https://doi.org/10.1016/j.conbuildmat.2016.03.013
41. Min KE, Jeong HM (2013) Characterization of air-blown asphalt/trans-polyoctenamer rubber blends. J Ind Eng Chem 19:645–649. https://doi.org/10.1016/j.jiec.2012.09.017
42. González V, Martínez-Boza FJ, Navarro FJ, Gallegos C, Pérez-Lepe A, Páez A (2010) Thermomechanical properties of bitumen modified with crumb tire rubber and polymeric additives. Fuel Process Technol 91:1033–1039. https://doi.org/10.1016/j.fuproc.2010.03.009
43. Khodaii A, Mehrara A (2009) Evaluation of permanent deformation of unmodified and SBS modified asphalt mixtures using dynamic creep test. Constr Build Mater 23:2586–2592. https://doi.org/10.1016/j.conbuildmat.2009.02.015
44. Qian C, Fan W, Ren F, Lv X, Xing B (2019) Influence of polyphosphoric acid (PPA) on properties of crumb rubber (CR) modified asphalt. Constr Build Mater 227:117094. https://doi.org/10.1016/j.conbuildmat.2019.117094
45. Baldino N, Gabriele D, Rossi CO, Seta L, Lupi FR, Caputo P (2012) Low temperature rheology of polyphosphoric acid (PPA) added bitumen. Constr Build Mater 36:592–596. https://doi.org/10.1016/j.conbuildmat.2012.06.011
46. EN 1426 (2015) Bitumen and bituminous binders—determination of needle penetration. European Committee for Standardization, Brussels, Belgium
47. EN1427 (2015) Bitumen and bituminous binders—determination of the softening point—ring and Ball method. European Committee for Standardization, Brussels, Belgium
48. ASTM E168 (2016) Standard practices for general techniques of infrared quantitative analysis (FTIR). ASTM International, West Conshohocken, PA, USA
49. EN14770 (2012) Bitumen and bituminous binders—determination of complex shear modulus and phase angle—dynamic shear rheometer (DSR). European Committee for Standardization, Brussels, Belgium
50. Liu S, Cao W, Fang J, Shang S (2009) Variance analysis and performance evaluation of different crumb rubber modified (CRM) asphalt. Constr Build Mater 23:2701–2708. https://doi.org/10.1016/j.conbuildmat.2008.12.009
51. Airey GD, Rahman MM, Collop AC (2003) Absorption of bitumen into crumb rubber using the basket drainage method. Int J Pavement Eng 4:105–119. https://doi.org/10.1080/1029843032000158879
52. Rodríguez-Fernández I, Tarpoudi Baheri F, Cavalli MC, Poulikakos LD, Bueno M (2020) Microstructure analysis and mechanical performance of crumb rubber modified asphalt concrete using the dry process. Constr Build Mater 259:119662. https://doi.org/10.1016/j.conbuildmat.2020.119662
53. Liao MC, Lo TJ (2021) Material characterization and balanced design of asphalt-rubber binders. J Mater Civ Eng 33:04020424. https://doi.org/10.1061/(ASCE)MT.1943-5533.0003504
54. ASTM D6114M-19 (2019) Standard specification for asphalt-rubber binder. ASTM International, West Conshohocken, PA, USA
55. Williams ML, Landel RF, Ferry JD (1955) The temperature dependence of relaxation mechanisms in amorphous polymers and other glass-forming liquids. J Am Chem Soc 77:3701–3707. https://doi.org/10.1021/ja01619a008

56. Airey GD (2002) Use of black diagrams to identify inconsistencies in rheological data. Road Mater Pavement Des 3:403–424. https://doi.org/10.1080/14680629.2002.9689933
57. Lo Presti D (2013) Recycled tyre rubber modified bitumens for road asphalt mixtures: a literature review. Constr Build Mater 49:863–881. https://doi.org/10.1016/j.conbuildmat.2013.09.007
58. Lo Presti D, Airey G (2013) Tyre rubber-modified bitumens development: the effect of varying processing conditions. Road Mater Pavement Des 14:888–900. https://doi.org/10.1080/14680629.2013.837837
59. Rodríguez-Alloza AM, Gallego J, Giuliani F (2016) Complex shear modulus and phase angle of crumb rubber modified binders containing organic warm mix asphalt additives. Mater Struct 50:77. https://doi.org/10.1617/s11527-016-0950-1

Chapter 4
Waste Aggregates in Asphalt Mixtures

Emiliano Pasquini, Fernando Moreno-Navarro, Augusto Cannone Falchetto, Marco Pasetto, Giovanni Giacomello, Raul Tauste-Martinez, Di Wang, Michel Vaillancourt, Alan Carter, Éric Lachance-Tremblay, Nunzio Viscione, Francesca Russo, Marta Skaf, Marko Orešković, Ana Cristina Freire, David Hernando, Peter Mikhailenko, and Lily D. Poulikakos

Abstract As part of the RILEM Technical Committee TC 279-WMR on valorisation of waste and secondary materials for roads, Task Group 3 used various waste aggregates as a replacement for natural aggregates in asphalt mixtures, namely, construction and demolition waste, recycled concrete aggregates and steel slag. For this interlaboratory exchange program, the mixtures were produced by substituting various amounts of virgin aggregates with the aforementioned waste aggregates in various fractions. The results from eight laboratories were compared to a reference mixture in each laboratory with conventional aggregates. The mechanical performance of these mixtures using such alternative aggregates indicates that bituminous mixtures prepared with the investigated waste aggregates required higher amounts of bitumen but had acceptable volumetric properties. Furthermore, they had better mechanical performance, regardless of the source of recycled aggregates. The results

E. Pasquini (✉) · M. Pasetto · G. Giacomello
Department of Civil, Environmental and Architectural Engineering (ICEA), University of Padova, Via Marzolo, 9-35131 Padova, Italy
e-mail: emiliano.pasquini@unipd.it

F. Moreno-Navarro · R. Tauste-Martinez
LABIC-UGR, Laboratorio de Ingeniería de La Construcción Universidad de Granada ETS Caminos, Canales y Puertos Severo Ochoa S/N Campus de Fuentenueva, 18071 Granada, Spain

A. Cannone Falchetto · D. Wang
Department of Civil Engineering, Aalto University, Rakentajanaukio, 4-02150 Espoo, Finland

M. Vaillancourt · A. Carter · É. Lachance-Tremblay
Department of Construction Engineering, École de Technologie Supérieure (ETS), Montreal, QC H3C, Canada

N. Viscione · F. Russo
Department of Civil, Construction and Environmental Engineering, University of Naples Federico II, Naples, Italy

M. Skaf
Department of Construction, Escuela Politécnica Superior, University of Burgos, C/Villadiego s/n, 09006 Burgos, Spain

© The Author(s), under exclusive license to Springer Nature Switzerland AG 2023

A. Cannone Falchetto et al. (eds.), *Valorisation of Waste and Secondary Materials for Roads*, RILEM State-of-the-Art Reports 38,
https://doi.org/10.1007/978-3-031-33173-2_4

show that such aggregates can be a viable option for use in roads, thereby contributing to the zero-waste society and reducing the use of natural aggregates. However, the adaptation of quality control/acceptance criteria and pavement design as well as the development of specific research aimed at assessing the environmental viability and life cycle sustainability of such materials is necessary.

Keywords Asphalt mixture · Recycling · Steel slag · Recycled concrete aggregates · Construction and demolition waste · Interlaboratory · Mix design · Mechanical characterisation

1 Introduction

The growing demand for building more sustainable pavements within the rising principles of the circular economy includes different aspects and components of the road infrastructures, from the construction processes to the selection and design of the materials [1–3]. Asphalt mixture is an essential element of flexible pavements. It consists of a mix of aggregates/filler (90–95% of the mixture by weight and 75–85% by volume), bitumen (5–10% by mixture weight and 10–20% by volume), and air (4–7% by volume). Therefore, aggregates represent an essential constituent; commonly, they are obtained from quarries and mining operations and can be further processed. These activities are highly material and energy-intensive [4, 5] due to the exploitation of non-renewable resources and the associated release of pollutants, with a negative environmental impact [6].

Over the years, to minimise the exploitation of natural materials and improve resource efficiency, various secondary and marginal materials obtained from different activities and processes have been considered to replace conventional mineral aggregates for paving mixtures [7–9]. These alternatives include various by-products such as Steel Slags (SS), Construction and Demolition Waste (CDW), Recycled Concrete Aggregates (RCA), crumb rubber, Recycled Asphalt Pavements (RAP), recycled asphalt shingles (RAS), waste plastics, and polymers. Among these marginal materials, SS, CDW and RCA are mainly used as replacements for the natural aggregates,

M. Orešković
University of Belgrade, Faculty of Civil Engineering, Bulevar Kralja Aleksandra 73, 11000 Belgrade, Serbia

A. C. Freire
Transportation Department, National Laboratory for Civil Engineering (LNEC), Avenida do Brasil, 101, 1700-066 Lisboa, Portugal

D. Hernando
Sustainable Pavements and Asphalt Research (SuPAR), University of Antwerp, Antwerp, Belgium

P. Mikhailenko · L. D. Poulikakos
Empa, Swiss Federal Laboratories for Materials Science and Technology, Überlandstrasse 129, 8600 Dübendorf, Switzerland

while the others can also be used as binder modifiers/enhancers. Regarding the use of alternative materials, Piao et al. have shown that some of these materials can reach the performance of all virgin materials but have reached a varied technology readiness level (TRL). For example, they estimated the TRL for SS to be 7–9, indicating that the application is partially or entirely industrialised, whereas the TRL of RCA to be 1–4, indicating the progress at the laboratory scale or lower [10].

Due to their advantageous mechanical properties, steel slag (SS) has been used in the road industry for decades [11]. Different types of SS, granulated slag (BF) [12], Basic Oxygen steel slag (BOSS) [13], Electric Arc Furnace steel slags (EAFSS) [14, 15], and Linz-Donawitz (LD) slag [16], have been investigated in the laboratory environment. Previous studies showed that asphalt mixtures partially or entirely prepared with slags exhibit comparable performance properties with equivalent mixes designed with conventional natural aggregates. In most cases, better performances were observed in slags containing mixtures due to their high stiffness and adhesion properties. Such characteristics are observed at different material scales, such as mixture, mastic, and mortar [17, 18]. Beside the additional bitumen absorption, the swelling of steel slag restricted their application at a high Technology Readiness Level (TRL) in the past. Nowadays, in several countries, steel slags with a limited expansion rate can be accepted for asphalt pavement construction [19, 20]. In the case of Construction and Demolition Waste (CDW), the annual production in Europe reached 461 million tonnes, which causes approximately 30% of all waste generated [21]. In addition, such materials are responsible for one of the heaviest and most voluminous waste streams in the EU [22]. Hence, CDW has been identified by the EU as a priority waste stream for reuse and recycling, and a minimum recycling rate of 70% has been set for non-hazardous CDW by 2020 [3]. Several studies explored the use of CDW as an alternative aggregate in asphalt mixture. However, CDW consists of different materials, varying in content: 12–40% concrete, 8–54% masonry, 4–26% asphalt, 2–4% wood, and a small amount of glass [23]. Such components may change due to the location, origin, and batches. In addition, some of the materials contained in the CDW may not be suitable for asphalt mixtures. The use of up to 20% CDW in asphalt mixtures has shown similar performance properties compared to conventional mixtures [24]. However, incorporating high amounts of CDW in asphalt mixture is limited due to its variability and the presence of undesirable materials [25].

Cement concrete is the most used construction material in the civil engineering sector, with 25 billion tons of new products and 900 million tons of Recycled Concrete Aggregate (RCA) waste annually [26]. The reuse of RCA in different types of constructions, including asphalt pavement, has been studied for several decades [27–29]. Coarse and fine aggregate, together with fillers, have been incorporated in different layers, and the possible application of up to 100% RCA replacement in aggregates for asphalt mixture has been investigated [30, 31].

Previous studies validated that the use of high content of SS, CDW, and RCA in the asphalt mixture design could lead to comparable and even better performance properties [13, 32–34]. However, most of these studies were restricted to a single source of secondary aggregate, hindering a substantial comparison among the valuable results of these investigations. Moreover, consistently recycling such materials into asphalt mixtures would have multiple benefits: reduce the amount of waste, save virgin aggregates, diminish the production costs, promote conscious and virtuous economic and environmental practices, and limit the impact of the paving construction industry on the environment and society.

2 Interlaboratory Test Program

2.1 *Introduction and Objective*

Given the above background, a specific Task Group (TG3) of the RILEM Technical Committee (TC) 279-WMR on "Valorisation of Waste and Secondary Materials for Roads" was established to investigate through interlaboratory investigations the feasibility of replacing natural virgin aggregates with different types of by-products, such as Steel Slag (SS), Construction and Demolition Waste (CDW), and Recycled Concrete Aggregate (RCA).

In particular, TC 279-WMR was organised in four TGs, which focused on the possible use of waste polyethylene, crumb rubber from end-of-life tires and secondary aggregates in asphalt binders and mixtures. The work of the TC consisted of interlaboratory experimental test campaigns as well as relevant life cycle assessments. These activities actively involved more than twenty laboratories worldwide. In this regard, 17 members from nine European laboratories along with one laboratory from Canada contributed to the interlaboratory research activities launched by TG3 on "Waste Aggregates in Asphalt Mixtures".

Such experimental research was based on comprehensive laboratory characterisation of marginal aggregates and related mixtures, asking the participating members to collect materials in their own countries. This approach avoided the issues related to shipping a massive amount of materials worldwide while allowing a representative picture of each local condition.

The experimental activities were organised in: (i) a preliminary characterisation of the selected aggregates, followed by (ii) mix design and (iii) performance characterisation of the asphalt mixtures prepared with such secondary aggregates.

2.2 *Experimental Activities*

The waste aggregates were characterised in terms of gradation, particle density, water absorption, resistance to fragmentation, quality of fines and shape properties according to the relevant European and American standards. As discussed in the following analysis of results, all the laboratories used limestone aggregates as reference virgin material except for Lab8, which selected sandstone for that purpose.

Furthermore, only Lab1 studied SS, CDW and RCA, whereas the other laboratories provided contributions for only one or two of such secondary aggregates.

Laboratories were asked to carry out the mix design of asphalt mixtures containing one of the selected by-products or a combination of them. The study of a corresponding reference material prepared with only virgin aggregates was also recommended in order to highlight the contribution of the alternative material for comparison purposes. The mixes were prepared with raw materials obtained by each participant in their own country; in addition, a common type of virgin aggregate and bitumen were identified, i.e., limestone (L) virgin (V) aggregate and 50/70 penetration grade bitumen. All the laboratories complied with these recommendations except for Lab8, which used sandstone (S) aggregates, as this is the usual aggregate used in that country. Moreover, TG3 Leaders delivered detailed guidelines to the participants in order to homogenise the procedures among the laboratories. To this aim, the Marshall mix design procedure was suggested to identify the mix recipe for an AC16 (i.e., asphalt concrete with a 16 mm maximum aggregate size) mixture generally used for a binder course. In particular, the compacted specimens prepared through an impact (i.e., Marshall) compactor [35] by applying 75 blows per face should result in a target void content between 4 and 6% with a granulometric distribution falling within the band reported in Table 1. The optimum bitumen content should be identified based on mix design specifications typically applied in each country and included within the range detailed in Table 1. 0% (reference), 50% and 100% recycling rates by weight (filler excluded) were suggested even if different recycling rates were investigated when the properties of the marginal aggregate (dimensions, etc.) did not allow the above-mentioned recycling rates. Lab1 also investigated a couple of mixtures containing more than one secondary aggregate, as reported in Table 2. Here, detailed information about the aggregate and filler types and contents are reported for all the mixtures studied by the different laboratories, along with the range of the investigated binder content (BC). As can be seen, the mixtures were coded according to the types and content of the marginal aggregates used. Based on those compositions, Fig. 1 depicts the effective mix gradations adopted by the involved laboratories in terms of cumulative passing vs. sieve size.

It is worth noting that Lab3 and Lab4 produced AC10 and AC20 mixes, respectively, instead of the suggested AC16 due to the available aggregates and the prevalent experiences in their countries. Moreover, Lab3 only substituted part of the filler with CDW without replacing the coarser fractions. Finally, Lab8 provided results related to a special mixture, i.e., a Semi-Dense Asphalt with 4 mm maximum aggregate size (SDA4) prepared with a polymer-modified binder (PMB) and with RCA aggregates partially substituting the 2/4 mm fraction only resulting in 16% voids content. This depicted a wide and somehow heterogeneous picture at the interlaboratory level anyway, gaining a broader impression of the effect of alternative aggregates on various mixtures. Furthermore, the preparation and testing of a corresponding reference mixture (i.e., not including any secondary aggregate) through the same procedures led to consistent "internal" comparison thus allowing obtaining generally valid indications about the contribution of the selected by-products.

Table 1 Target AC16 mix composition

Sieve size (mm)	Lower limit (% passing)	Upper limit (% passing)
32	100	100
16	90	100
10	73	85
4	45	56
2	28	38
0.5	16	24
0.25	11	18
0.063	4	8

*Bitumen content: 4–6% by weight of aggregates

Table 2 Detailed composition of the investigated mixtures [36]

Lab	Mixture code	Aggregate content (%)				Filler		BC
		V	SS	CDW	RCA	Content (%)	Type	(%)
1	REF	95				5	L	4.0–6.0
	50SS	48	48			4	L	4.0–6.0
	100SS		92			8	L	4.0–6.0
	50SS50CDW		47	47		6	L	6.0–8.0
	50SS50RCA		45		45	10	L	5.0–7.0
2	REF	93				7	L	4.1–4.7
	25RCA	68			25	7	L	4.4–5.0
	50RCA	43			50	7	L	4.5–5.2
3	REF*	96				4	L	4.4–6.0
	8CDW*, **	92		8		–	L/CDW	4.4–6.0
	16CDW*, **	84		16		–	CDW	5.2–6.4
4	75SS***	17	75			8	L	3.0–5.0
6	REF	90				10	L	4.0–4.5
	50SS	46.5	46.5			7	L	4.5–5.25
	50CDW	45.5		45.5		9	L	4.5–5.25
	100SS		92			8	L	5.0–6.0
	100CDW			93		7	L	4.5–5.5
7	REF	100				–	L	4.5–5.5
	50CDW	46		50		4	L	4.4–6.0
	85CDW	14		81		5	L	5.5–6.5
8	REF****	92				8	S	6.0
	35RCA (2/4)****	59			33	8	S	6.0
	70RCA (2/4)****	25			67	8	S	6.4

*Only filler substitution; **AC10; ***AC20; ****SDA4 with Polymer modified binder (PMB) and voids content of 16%

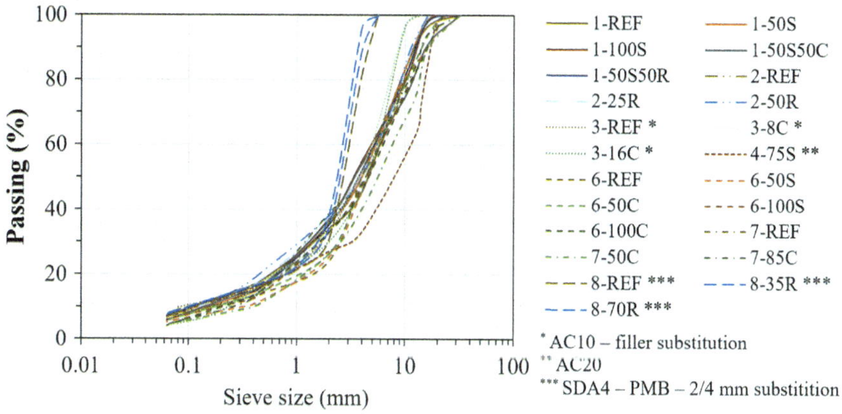

Fig. 1 Gradation of the investigated mixtures

Finally, the optimised mixtures were subjected to a basic performance characterisation study aimed at investigating the effect of the different waste types and contents on volumetrics, strength, stiffness, water resistance and anti-rutting potential.

To this aim, testing specimens were prepared using the Marshall compactor (75 blows per face) using the above-mentioned raw materials with the bitumen dosed according to the optimum binder content (OBC). Then, the volumetric properties, with particular reference to the air void content (AVC), were measured according to EN 12697-8 [37], whereas the indirect tensile strength (ITS) at 25 °C was determined through static tests carried out under displacement control (speed rate equal to 51 mm/min) along the vertical diameter according to EN 12697-23 [38].

Analogous ITS tests carried out on similar specimens subjected to a specific moisture conditioning allowed to assess the water resistance of the investigated materials in terms of indirect tensile strength ratio (ITSR), in percent. Moisture conditioning was carried out according to EN 12697-12 [39] by placing the tested specimens in a water bath kept at 40 °C for 72 h. ITS tests were then executed at 25 °C after the prescribed 4-h of conditioning.

Some laboratories also measured the stiffness characteristics of the investigated materials following the procedure described in EN 12697-26/Annex C [40]. In particular, the stiffness was determined on cylindrical specimens subjected to haversine pulses along the vertical diameter in indirect tensile strength configuration (ITSM). Tests were carried out in strain-controlled mode by setting the target horizontal deformation equal to 5 μm with a rise time (i.e., the time needed to reach the peak load) equal to 124 ms.

Furthermore, five laboratories provided contributions concerning the permanent deformation resistance of asphalt mixtures prepared with marginal aggregates. Two labs carried out the suggested uniaxial cyclic compression tests (CCT) with confinement according to method A1 described in EN 12697-25 [41]. In particular, 150 mm diameter cylindrical specimens prepared with a gyratory compactor (EN 12697-31

Table 3 Performance characterisation interlaboratory test campaign

Laboratory	Test				
	ITS	ITSR	ITSM	CCT	WTT
1	×	×	×	×	
2	×	×	×		×
3	×	×			×
4	×	×			×
6	×	×	×	×	
7	×	×			
8	×	×			

[42]) were subjected to cyclic axial load pulses in stress-controlled configuration. Tests were executed at 40 °C and 100 kPa stress level transmitted through a centred 100 mm diameter upper loading plate; 3,600 loading cycles were applied so that the specimen was subjected to 1 s loading time and 1 s rest period (i.e., test frequency of 0.5 Hz). Mixture performance was evaluated in terms of creep rate (i.e., the slope of the quasi-linear part of the evolution of the cumulative axial strain as a function of the number of loading cycles) and final strain (i.e., the cumulative axial strain at the end of the test). Additionally, three laboratories investigated the rutting potential of the studied materials through wheel tracking tests (WTT) carried out in air at 60 °C according to EN 12697-22 [43]. To this aim, slabs were prepared using a roller compactor compliant with EN 12697-33 [44]. Then, the specimens were subjected to a repeated passing of a loaded pneumatic wheel moving at a fixed frequency (i.e., passes per minute). Since the three laboratories adopted different test conditions (i.e., number and frequency of passes, applied load, etc.), results were analysed in terms of final strain of the waste sample compared to the corresponding reference sample. Table 3 summarises the performance characterisation tests carried out by the participating laboratories.

3 Results and Discussion

3.1 Aggregate Characterisation

Before the mix design procedure, virgin and secondary aggregate materials were initially characterised. Geometrical, physical and mechanical properties, including particle density (PD), water absorption (WA), Los Angeles (LA), sand equivalent (SE), flakiness index (FI), and shape index (SI) were determined, and testing results are given in Table 4. Additionally, Lab4 determined micro-Deval (M_{DE}) and methylene blue (MB) values of the used SS, whereas the same laboratory did not provide properties of the virgin aggregate. It should be emphasised that all laboratories used the virgin (reference) aggregate of a limestone origin, except for Lab8, which used sandstone to produce SDA4 mixture.

The testing results of the interlaboratory study showed that all limestone aggregates have comparable properties regardless of their origin from all over the world. Since the water absorption (WA) of all aggregates is only of the order 0.3–1.4%, the values of all virgin aggregates' particle densities (PD) are very close to each other, in the range between 2.66 and 2.80 Mg/m^3. Limestone resistance to fragmentation, expressed through the Los Angeles coefficient (LA), was very uniform, all having values between 20 and 25, whereas the sandstone aggregate had a much lower LA value of 13. Only two laboratories reported flakiness (FI) and shape indexes (SI), and their average values were the same, 9%. The biggest variation in the testing results of limestone aggregate was in the case of sand equivalent (SE), which varied between 63 and 95%, with an average value of 76%, and sandstone somewhat lower at 54%.

Table 4 Basic properties of aggregates used in the study [36]

Lab	Aggregate	Properties					
		PD	WA	LA	SE	FI	SI
		(Mg/m^3)	(%)	(–)	(%)	(–)	(–)
		EN 1097-6 AASHTO T 19 M/T 19		EN 1097-2 ASTM C131/ 131 M	EN 933-8 ASTM T 176	EN 933-3	EN 933-4
1	Limestone	2.66	0.75	21	63	8	12
	SS	3.71	1.02	12	85	6	9
	CDW	2.56	2.40	36	35	8	18
	RCA	2.45	1.77	28	68	10	18
2	Limestone	2.69	0.30	26	–	–	–
	RCA	2.43	3.55	32	92	–	–
3	Limestone	2.71	0.69	–	–	–	–
	CDW	1.54	23.1	–	39	–	–
4	SS	3.61	0.73	≤15	87	≤10	≤15
5	Limestone	2.70	1.37	23	68	–	–
	SS	3.39	2.09	22	98	–	–
	RCA	2.65	6.81	35	82	–	–
6	Limestone	2.80	0.69	25	77	–	–
	SS	3.34	3.19	20	70	–	–
	CDW	2.77	5.03	33	60	–	–
7	Limestone	2.71	-	20	95	10	6
	CDW	2.63	-	41	95	13	21
8	Sandstone	2.71[a]	0.73[a]	13	54[c]	–	–
	RCA	2.63[a]	4.26[a]	26[b]	89[c]	–	–

[a]2/4 mm aggregate size; [b]4/8 mm aggregate size; [c]0.1/2 mm aggregate size

All steel slag aggregates (SS) used in this study had a PD of roughly 30% higher in comparison with virgin aggregate (between 3.34 and 3.71 Mg/m^3), with a better fragmentation resistance (LA coefficient lies between 12 and 22). The porous structure of the steel slag grains in some cases led to higher WA in comparison to limestone aggregate, but it should also be mentioned that two laboratories used material with WA sufficiently low (<2%) to be considered a freeze/thaw resistant material according to EN 13043 [45]. Average SE value was generally above the limestone values for the SS, because of the lower content of fines, whereas FI and SI had comparable values with the virgin aggregates since the applied crushing process, similar to the one applied to natural aggregates, allowed obtaining a good particle shape.

As expected, construction and demolition waste (CDW) showed the worst characteristics among all materials used in this study. PD and SE values were much lower, and WA and LA were much higher than all other materials. In addition to the significantly worse properties, it is necessary to highlight that CDW showed a noticeable variability among all participating laboratories. This may be a consequence of the heterogeneity of this type of material due to the methodology applied during the demolition that leads to the presence of different waste materials within waste aggregate: wood, glass, soil, ceramic, concrete, bitumen, etc. However, extremely high-water absorption (2.4–23.1%) and low resistance to abrasion (LA = 33–41) may be two limiting factors for the application of CDW in new asphalt mixtures when considering existing technical specifications for asphalt mixtures. SE value is also important in the manufacturing process of new asphalt mixtures because unsatisfactory low results may lead to reduced binder-aggregate bond. The application of CDW can be validated through its application on real test sections, followed by the mechanical evaluation.

In this study SE value also varied a lot—from 35 to 95—that should be considered when using this kind of material and furthermore, the importance of its appropriate handling.

Finally, RCA materials mostly had poorer properties than limestone and sandstone aggregates and SS but were slightly better than CDW. Only one lab reported using RCA with relatively low water absorption (WA = 1.77%), whereas others reported much higher WA values, going up to 4.26–6.81%. All RCA materials had acceptable SE values (68–89%), unlike LA values, which were relatively uniform (28–35%) but still higher than standard requirements. The shape indexes obtained for RCA-FI and SI—were of the same order of magnitude as those obtained for CDW and slightly higher than those of limestone.

It must be taken into consideration that the laboratory results obtained for the waste aggregate materials were compared to the usual limits applied to natural aggregates. So, it is to be expected that some parameters evaluated do not comply. Its incorporation in the production of bituminous mixtures and its assessment will validate its use.

Figure 2 presents the average testing results of Particle Density, Water Absorption, Los Angeles coefficient, and Sand Equivalent along with the respective minimum and maximum values measured by the different labs, allowing comparing the obtained results for the four studied aggregate materials. In addition, it is possible to observe

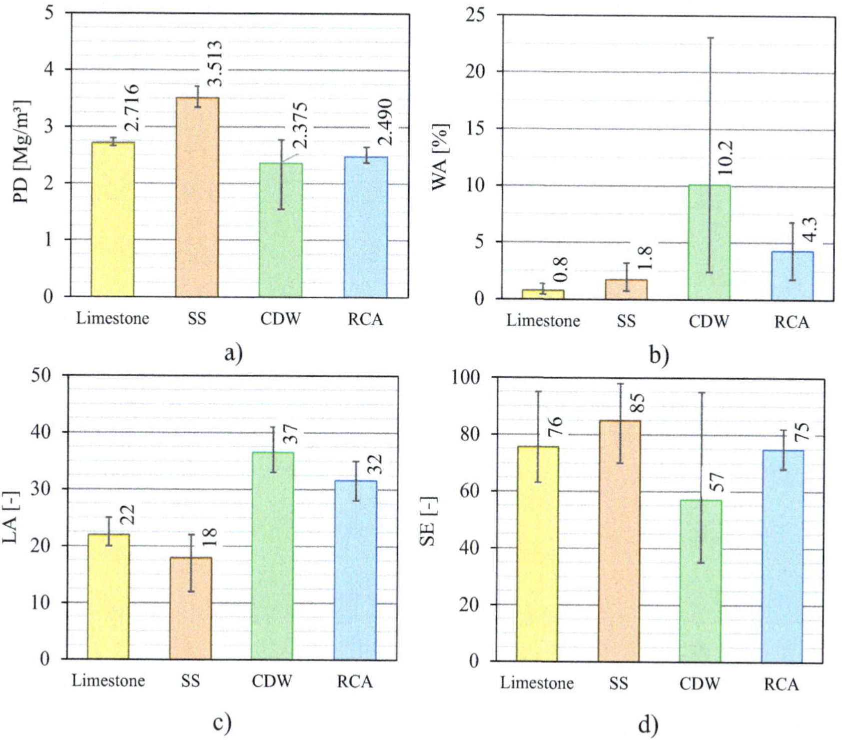

Fig. 2 Testing results of **a** particle density, **b** water absorption, **c** Los Angeles coefficient, and **d** sand equivalent

the greater variability of results obtained for CDW, namely for PD, WA and SE parameters.

3.2 Mix Design

The Marshall method was recommended to determine the optimum bitumen content (OBC) according to the target mix composition already reported in Table 1. For each mixture that was designed using waste materials, a reference mix (i.e., without waste aggregates) was considered by each laboratory involved for comparison purposes. Overall, Fig. 3 reports enhanced Marshall properties at the optimum binder content for asphalt mixtures containing waste aggregates; in particular, materials designed using CDW exhibited the highest mean value of Marshall Stability (MS) 26% and 7% more than reference and steel slag mixes, respectively (13.5 kN for CDW vs. 10.7 kN of REF and 12.6 kN of SS). In this sense, Fig. 4a depicts the average Marshal Stability measured at the OBC by the participating laboratories, regardless

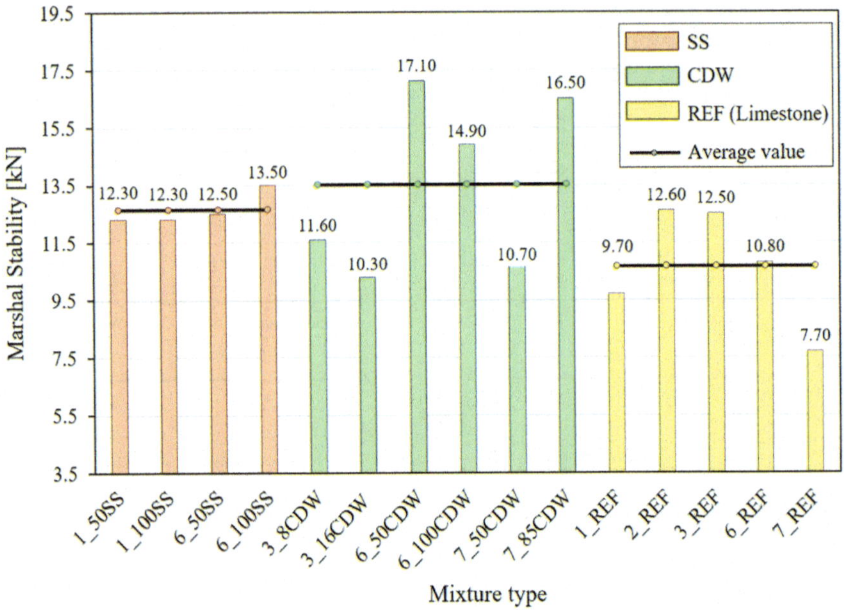

Fig. 3 Marshall Stability of optimised asphalt mixtures

of the amount of waste used, while only differentiating the type of waste. Each dot represents the average value of one mixture type from one lab. On the other hand, the same experimental findings are shown in Fig. 4b in terms of percentage increase of MS with respect to the corresponding reference mixture as a function of content of waste aggregates. Here, it is even more evident that the MS generally increases thanks to the use of secondary aggregates even if a less clear effect due to by-product type and content appears. In particular, high scattering was observed in the case of CDW, likely due to the heterogeneous characteristics of such type of aggregates.

On the other hand, Fig. 5 shows that the asphalt mixes designed with waste materials required higher OBC than the corresponding reference mixtures, with an average increase of about 18% (OBC is equal to 5.2% and 5.5% for SS and CDW materials, respectively vs. 4.5% of REF).

These experimental findings align with the water absorption of the secondary aggregates displayed in Fig. 2; generally, the higher the WA, the higher the OBC. This fact can be clearly observed in Fig. 6, which reports the increase in OBC with respect to the corresponding reference mixture as a function of by-product content; again, the different marginal aggregates types are characterised by different symbols and colours.

At the same time, Fig. 7 demonstrates that the increase of OBC in the case of secondary mixtures also allowed to guarantee acceptable volumetric characteristics with all the studied materials from each laboratory, meeting the 4–6% target in terms of air voids. However, it is worth noting that the asphalt mixes designed

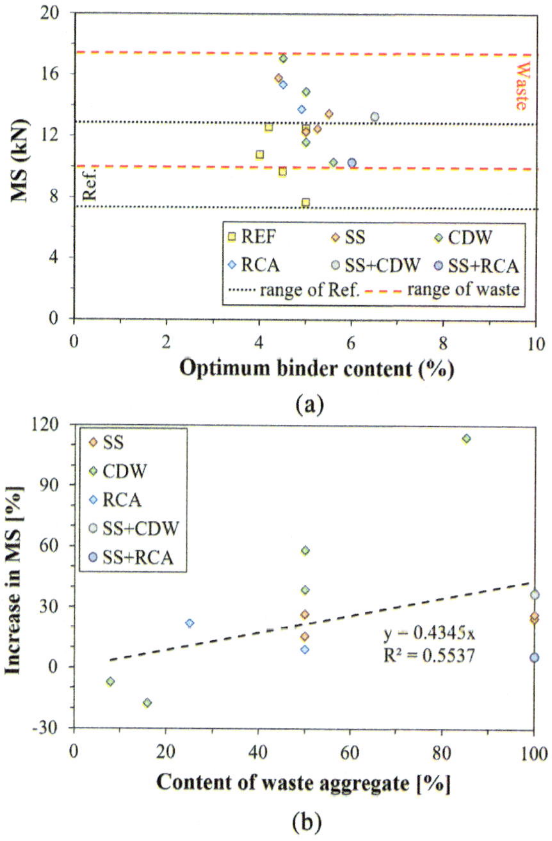

Fig. 4 **a** Average MS measured at the OBC by the different laboratories and **b** increase in MS with respect to the REF mixtures

using marginal materials displayed slightly higher air voids values than the reference mixture.

Since the asphalt mixtures designed with waste materials required a higher OBC, it was decided to perform a deeper volumetric characterisation to have a closer look at the binder absorbed by the aggregate. Figure 8 provides an example of three mixtures produced by different laboratories with different secondary materials. It is clear that the binder absorption of the mixtures with waste aggregate was greater than that of the corresponding reference mix, thus confirming the relationship between the higher amount of bitumen required and the type of aggregate used. In particular, the increase in absorption was relatively small for steel slag (0.1–0.2%) but notably large for CDW and RCA (2–4%). Overall, even if it is difficult to quantify the exact increase in binder absorption as a function of the percentage of marginal material, it is evident that the increased binder absorption of waste material is largely responsible for the increased binder content of this type of asphalt mixtures.

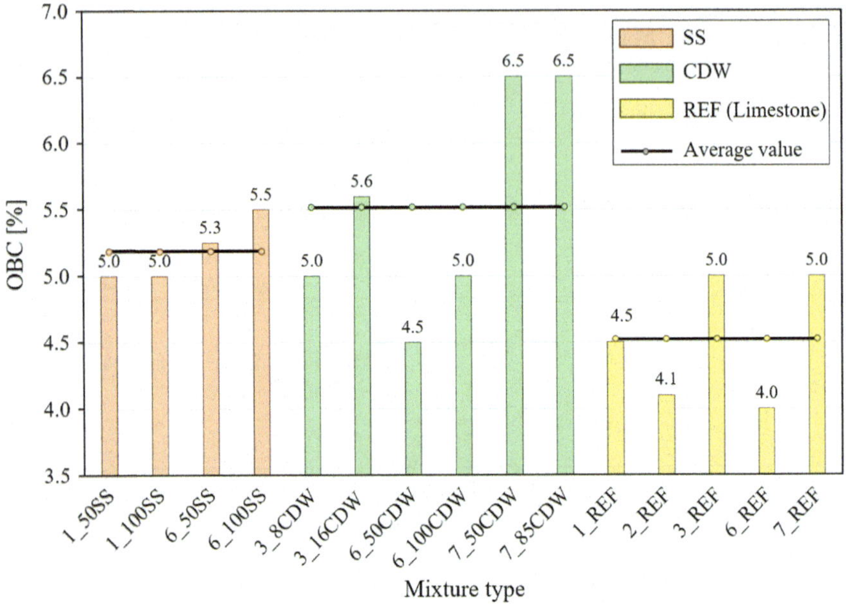

Fig. 5 Optimum bitumen content of studied asphalt mixtures

Fig. 6 Increase in OBC with respect to the corresponding REF mixtures

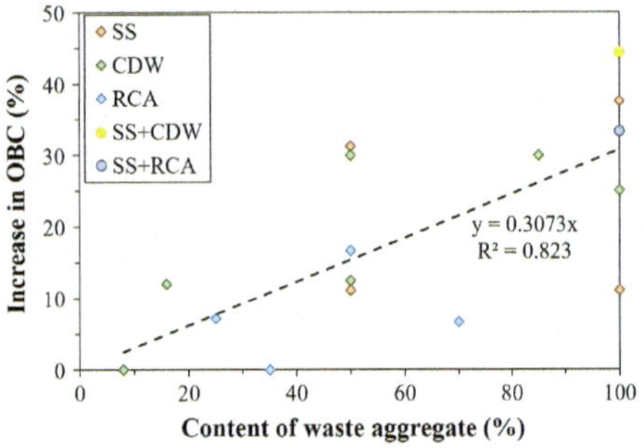

Fig. 7 Air voids content of designed asphalt mixtures

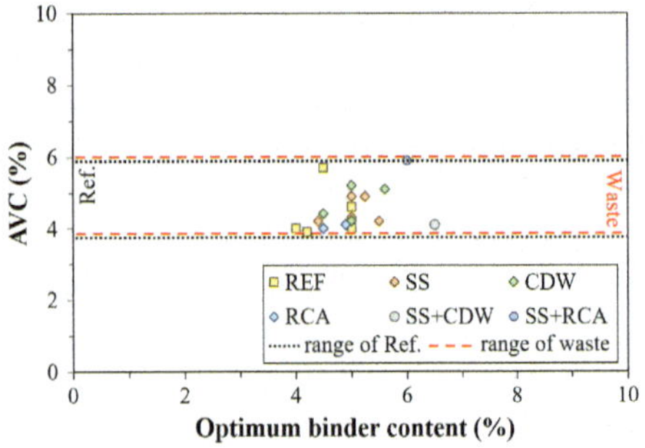

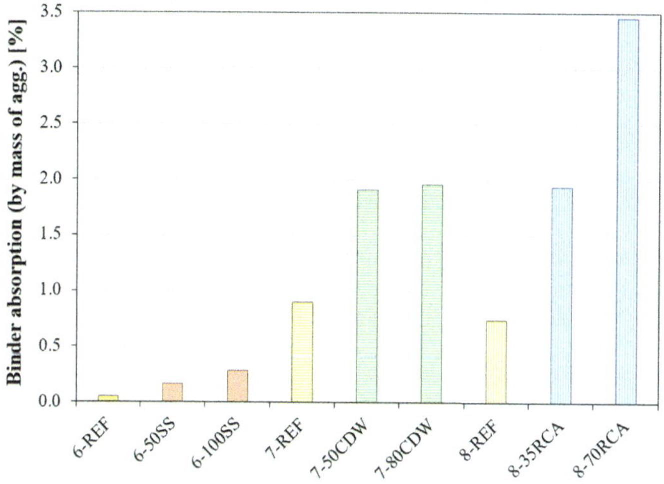

Fig. 8 Binder absorption for mixtures produced with different by-products

3.3 Mixture Characterisation

The results from the indirect tensile strength and indirect tensile strength ratio are shown in Fig. 9. Generally, the results show that waste aggregate addition has a beneficial effect on ITS (Fig. 9a). The %ITSR varied from approximately 60–105%, but no solid trends for a particular waste material could be seen (Fig. 9b). Except for 8_100RCA, all mixes met the acceptance criteria generally used (range of 70–90% depending on the specifications from the different countries).

In this sense, it is worth noting that all the materials prepared with waste aggregates were able to achieve satisfactory performance (i.e., acceptable by standard technical specifications for asphalt binder layers) in terms of both strength and moisture susceptibility, except for a couple of CDW mixtures (i.e., 3_8CDW and 8_100RCA) which experiences an excessive decrease in strength after water conditioning.

Promising results were also observed in the case of a combination of wastes (Lab. 1).

Four laboratories also investigated the permanent deformation resistance of waste mixes. The results for rutting/permanent deformation are shown in Fig. 10 and grouped by laboratory or waste aggregate. As mentioned in the mix design section, additional binder was added to all the mixtures to compensate for extra absorption by the secondary aggregates. While the results varied a lot for the materials with waste aggregates, the general tendency was that lower permanent deformation was observed for waste mixes even though binder content was higher.

In particular, Fig. 10b reporting the average results grouped by marginal mixtures along with the corresponding minimum and maximum values measured by the different labs shows that CDW was, on average beneficial to rutting resistance; the SS led to trends in both directions, while the RCA did not have a significant effect.

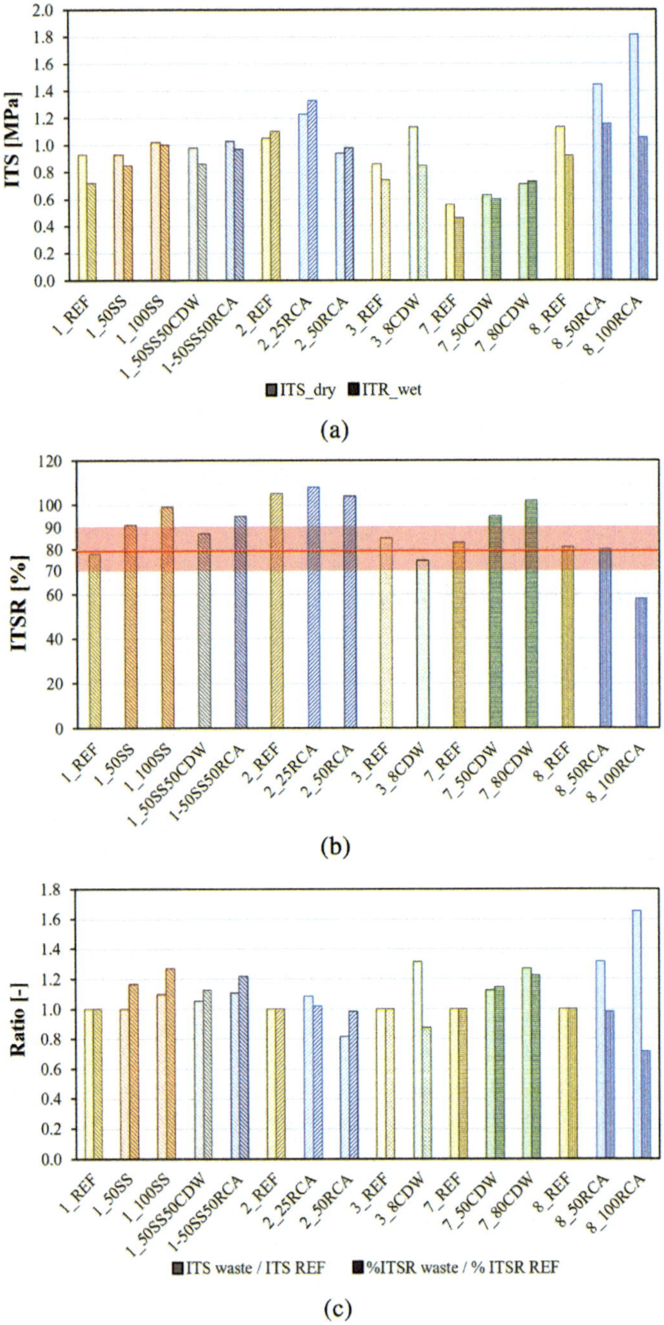

Fig. 9 **a** Indirect tensile strength (ITS) results by mixtures; **b** indirect tensile strength ratio (ITSR) results by mixture; **c** ITS and ITSR results' ratio of reference and waste mixture

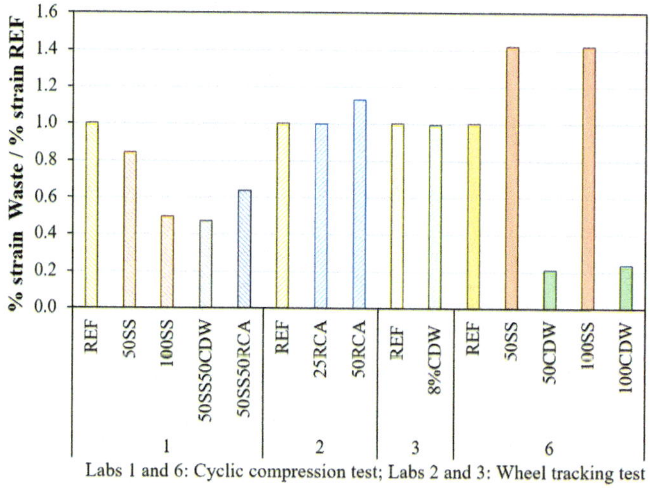

(a)

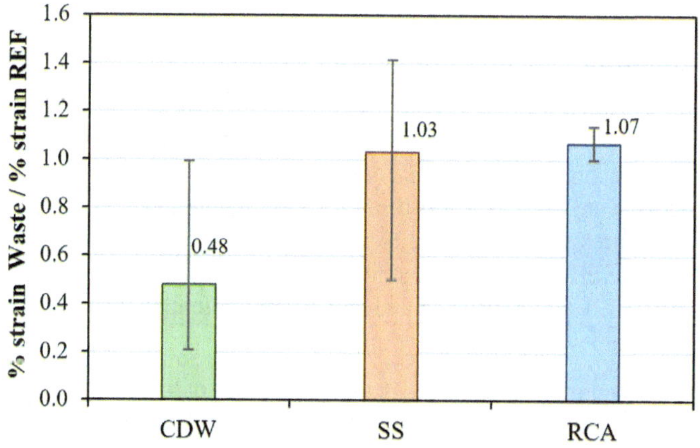

(b)

Fig. 10 Permanent deformation ratio: **a** by laboratory and, **b** by waste aggregate

Finally, the indirect tensile stiffness modulus (ITSM) results are shown in Fig. 11 as the ratio to the reference sample for the particular laboratory. Overall, the available experimental findings indicated a neutral to positive effect due to the waste aggregates, except for SS mixes tested by Lab6. This is in line with CCT test results from the same lab and could be ascribed to a possibly weaker SS aggregates source for that specific laboratory.

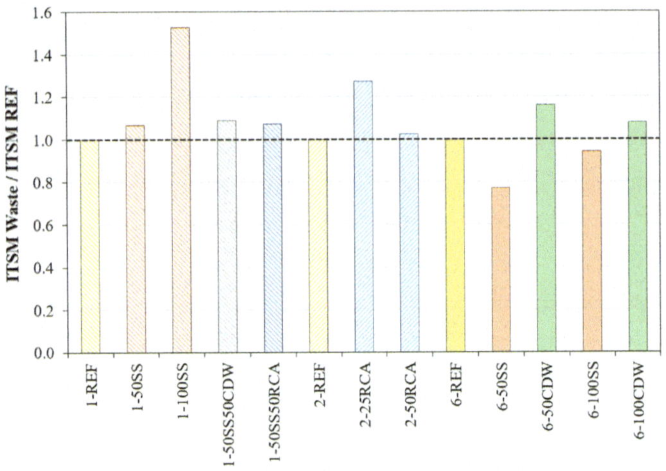

Fig. 11 ITSM results of waste mixtures relative to the reference

4 Conclusions

Under the framework of RILEM TC 279-WMR, various waste aggregates, namely recycled concrete aggregates (RCA), construction and demolition waste (CDW) and steel slag (SS), were used as a substitute for virgin aggregates in asphalt mixtures. Optimised mixtures were subjected to a basic performance characterisation study to investigate the effect of the different waste types and contents on volumetrics, strength, stiffness, water resistance and anti-rutting potential. The following can be concluded from this multi-laboratory investigation:

- All steel slag aggregates (SS) used in this study had a particle density (PD) roughly 30% higher than that of virgin aggregate, with a better fragmentation resistance.
- Construction and demolition waste (CDW) showed the worst characteristics among all materials used in this study. Particle density (PD) and sand equivalent (SE) values were much lower, and water absorption (WA) and Los Angeles abrasion loss (LA) much higher in comparison with all other waste and virgin materials. Extremely high-water absorption and low abrasion resistance may be two limiting factors for the application of CDW in new asphalt mixtures when considering existing technical specifications for asphalt mixtures.
- RCA materials mostly had poorer properties than limestone and sandstone aggregates and SS but slightly better than CDW
- The volumetric and mechanical investigation results show that more binder was required in most cases and the resulting mixtures have acceptable volumetric and Marshall stability properties. It was shown that there is a direct relationship between the higher amount of bitumen required and the type of aggregate used. In particular, the increase in absorption was relatively small for steel slag but notably large for CDW and RCA.

- The Marshall Stability (MS) values generally increased when waste aggregates were used.
- Waste aggregate addition did not have a detrimental effect on the indirect tensile strength (ITS).
- Except for one, all mixtures met the acceptance criteria generally used for water sensitivity; no solid trends for a particular waste material could be seen.
- The general trend was that lower permanent deformation was observed for mixtures with waste aggregates even though binder contents were higher.
- The results for the indirect tensile stiffness modulus (ITSM) indicated a positive effect due to the waste aggregates, except for SS mixtures tested by one laboratory.

Overall, the experimental findings collected at the interlaboratory level showed the need to adjust the mix design when alternative aggregates characterised by different particle densities and water absorption are used as partial or total substitution of the natural virgin aggregates. On the other hand, the optimised mixtures prepared with such secondary materials are able to guarantee equal or even better performance than those exhibited by the corresponding reference mixes. Furthermore, a notable variability of the characteristics was ascertained for each type of alternative aggregates used, not only due to different geographic locations.

5 Perspectives and Outlook

Over the years, to minimise the exploitation of natural materials and improve resource efficiency, various secondary and marginal materials obtained from different activities and processes have been considered to replace conventional mineral aggregates for paving mixtures. The work done by TC 279-WMR demonstrates that existing technical specifications for aggregates prescribe specific properties for water absorption and resistance to abrasion that are hard to meet for alternative aggregates. However, a majority of mixtures examined here fulfilled the performance criteria in terms of stiffness, permanent deformation resistance and moisture susceptivity while more research could be necessary about their cracking resistance (ITS provided promising results in this regard). Thus, one option for decision-makers would be to look at the performance of the mixture. At the same time, performance criteria that discriminate the weaknesses of using such alternative materials need to be developed. Furthermore, more field testing would provide a validation for the use of alternative aggregates. In order to increase the TRL level for such alternative aggregates, the following is recommended: (i) developing and widely disseminating the technical knowhow; (ii) investigating the environmental viability of such materials; (iii) performing life cycle assessments for the use of such waste streams in asphalt and comparison to existing processes; (iv) adjusting quality control criteria and adapting pavement design when alternative materials are used.

References

1. Van Dam TJ, Harvey J, Muench ST, Smith KD, Snyder MB, Al-Qadi IL, et al (2015) Towards sustainable pavement systems: a reference document (No. FHWA-HIF-15-002). Federal Highway Administration, USA
2. Pouranian MR, Shishehbor M (2019) Sustainability assessment of green asphalt mixtures: a review. Environments 6(6):73. https://doi.org/10.3390/environments6060073
3. Europe 2020 Strategy. https://ec.europa.eu/eurostat/web/europe-2020-indicators#:~:text=The%20Europe%202020%20strategy%20is,a%20sustainable%20social%20market%20economy. Accessed 18 Apr 2022
4. Ghanbari M, Abbasi AM, Ravanshadnia M (2018) Production of natural and recycled aggregates: the environmental impacts of energy consumption and CO_2 emissions. J Mater Cycles Waste Manage 20(2):810–822. https://doi.org/10.1007/s10163-017-0640-2
5. Zhang X, Su B, Yang J, Cong J (2019) Index decomposition and attribution analysis of aggregate energy intensity in Shanxi Province (2000–2015). J Cleaner Prod 238:117897. https://doi.org/10.1016/j.jclepro.2019.117897
6. Braga AM, Silvestre JD, de Brito J (2017) Compared environmental and economic impact from cradle to gate of concrete with natural and recycled coarse aggregates. J Cleaner Prod 162:529–543. https://doi.org/10.1016/j.jclepro.2017.06.057
7. Poulikakos LD, Papadaskalopoulou C, Hofko B, Gschösser F, Cannone Falchetto A, Bueno M et al (2017) Harvesting the unexplored potential of European waste materials for road construction. Resour Conserv Recycl 116:32–44. https://doi.org/10.1016/j.resconrec.2016.09.008
8. Rehman MU, Rashid K, Zafar I, Alqahtani FK, Khan MI (2020) Formulation and characterization of geopolymer and conventional lightweight green concrete by incorporating synthetic lightweight aggregate. J Build Eng 31:101363. https://doi.org/10.1016/j.jobe.2020.101363
9. Office JE, Chen J, Dan H, Ding Y, Gao Y, Guo M et al (2021) New innovations in pavement materials and engineering: a review on pavement engineering research 2021. J Traffic Transp Eng 8(6):815–999. https://doi.org/10.1016/j.jtte.2021.10.001
10. Piao Z, Mikhailenko P, Kakar MR, Bueno M, Hellweg S, Poulikakos LD (2021) Urban mining for asphalt pavements: a review. J Cleaner Prod 280:124916. https://doi.org/10.1016/j.jclepro.2020.124916
11. Yi H, Xu G, Cheng H, Wang J, Wan Y, Chen H (2012) An overview of utilization of steel slag. Proc Environ Sci 16:791–801. https://doi.org/10.1016/j.proenv.2012.10.108
12. Dunster AM (2002) The use of blast furnace slag and steel slag as aggregates. In: Proceedings of the 4th European Symposium on Performance of Bituminousand Hydraulic Materials in Pavements, Nottingham
13. Wu S, Xue Y, Ye Q, Chen Y (2007) Utilization of steel slag as aggregates for stone mastic asphalt (SMA) mixtures. Build Environ 42(7):2580–2585. https://doi.org/10.1016/j.buildenv.2006.06.008
14. Morone M, Costa G, Polettini A, Pomi R, Baciocchi R (2014) Valorization of steel slag by a combined carbonation and granulation treatment. Miner Eng 59:82–90. https://doi.org/10.1016/j.mineng.2013.08.009
15. Pasetto M, Baliello A, Giacomello G, Pasquini E (2017) Sustainable solutions for road pavements: a multi-scale characterization of warm mix asphalts containing steel slags. J Cleaner Prod 166:835–843. https://doi.org/10.1016/j.jclepro.2017.07.212
16. Groenniger J, Cannone Falchetto A, Isailović I, Wang D, Wistuba MP (2017) Experimental investigation of asphalt mixture containing Linz-Donawitz steel slag. J Traffic Transp Eng 4(4):372–379. https://doi.org/10.1016/j.jtte.2017.05.009
17. Moon KH, Cannone Falchetto A, Wang D, Riccardi C, Wistuba MP (2017) Mechanical performance of asphalt mortar containing hydrated lime and EAFSS at low and high temperatures. Materials 10(7):743. https://doi.org/10.3390/ma10070743

18. Pasetto M, Baliello A, Pasquini E, Skaf M, Ortega-López V (2020) Performance-based characterization of bituminous mortars prepared with ladle furnace steel slag. Sustainability 12(5):1777. https://doi.org/10.3390/su12051777
19. Skaf M, Manso JM, Aragón Á, Fuente-Alonso JA, Ortega-López V (2017) EAF slag in asphalt mixes: a brief review of its possible re-use. Resour Conserv Recycl 120:176–185. https://doi.org/10.1016/j.resconrec.2016.12.009
20. Freire AC, Maia A (2019) Estado da arte ibero-brasileira sobre a aplicação de agregados de escórias de aciaria em misturas betuminosas. Geotecnia (146):19–50. https://doi.org/10.24849/j.geot.2019.146.02 [in Portuguese]
21. European Commission, Construction and Demolition Waste (CDW). http://ec.europa.eu/environment/waste/construction_demolition.html. Accessed 18 Apr 2022
22. Construction and Demolition Waste, ERA. https://era.org.mt/topic/construction-and-demolition-waste/. Accessed 18 Apr 2022
23. Fischer C, Werge M (2009) Europe as a recycling society, present recycling levels of municipal waste and construction & demolition waste in the EU. Denmark, Copenhagen
24. Ossa A, García JL, Botero E (2016) Use of recycled construction and demolition waste (CDW) aggregates: a sustainable alternative for the pavement construction industry. J Cleaner Prod 135:379–386. https://doi.org/10.1016/j.jclepro.2016.06.088
25. Mikhailenko P, Arraigada M, Piao Z, Poulikakos LD (2022) Multiscale laboratory mechanical performance of SDA mixtures with construction and demolition waste filler. J Mater Civ Eng 34:04022106. https://doi.org/10.1061/(ASCE)MT.1943-5533.0004244
26. Lukasik J, Damtoft JS, Herfort D, Sorrentino D, Gartner EM (2007) Sustainable development and climate change initiatives. In: 12th International congress on the cement chemistry, Montreal, ST3, Canada
27. Acosta Alvarez D, Alonso Aenlle A, Tenza-Abril AJ, Ivorra S (2019) Influence of partial coarse fraction substitution of natural aggregate by recycled concrete aggregate in hot asphalt mixtures. Sustainability 12(1):250. https://doi.org/10.3390/su12010250
28. Radević A, Đureković A, Zakić D, Mladenović G (2017) Effects of recycled concrete aggregate on stiffness and rutting resistance of asphalt concrete. Constr Build Mater 136:386–393. https://doi.org/10.1016/j.conbuildmat.2017.01.043
29. Radević A, Isailović I, Wistuba MP, Zakić D, Orešković M, Mladenović G (2020) The impact of recycled concrete aggregate on the stiffness, fatigue, and low-temperature performance of asphalt mixtures for road construction. Sustainability 12(10):3949. https://doi.org/10.3390/su12103949
30. Chen MJ, Wong YD (2013) Porous asphalt mixture with 100% recycled concrete aggregate. Road Mater Pavement Des 14(4):921–932. https://doi.org/10.1080/14680629.2013.837839
31. Nwakaire CM, Yap SP, Yuen CW, Onn CC, Koting S, Babalghaith AM (2020) Laboratory study on recycled concrete aggregate based asphalt mixtures for sustainable flexible pavement surfacing. J Cleaner Prod 262:121462. https://doi.org/10.1016/j.jclepro.2020.121462
32. Gedik A (2020) A review on the evaluation of the potential utilization of construction and demolition waste in hot mix asphalt pavements. Resour Conserv Recycl 161:104956. https://doi.org/10.1016/j.resconrec.2020.104956
33. Xu X, Luo Y, Sreeram A, Wu Q, Chen G, Cheng S et al (2022) Potential use of recycled concrete aggregate (RCA) for sustainable asphalt pavements of the future: a state-of-the-art review. J Cleaner Prod 130893. https://doi.org/10.1016/j.jclepro.2022.130893
34. Arabani M, Moghadas Nejad F, Azarhoosh AR (2013) Laboratory evaluation of recycled waste concrete into asphalt mixtures. Int J Pavement Eng 14(6):531–539. https://doi.org/10.1080/10298436.2012.747685
35. EN 12697-30 (2018). Bituminous mixtures—test methods—Part 30: specimen preparation by impact compactor. European Committee for Standardization, Brussels, Belgium
36. Pasetto M, Pasquini E, Giacomello G, Moreno-Navarro F, Tauste-Martinez R, Cannone Falchetto A, Vaillancourt M, Carter A, Viscione N, Russo F, Skaf M, Oreskovic M, Freire AC, Mikhailenko P, Poulikakos LD (2022) An interlaboratory test program on the extensive use of waste aggregates in asphalt mixtures: preliminary steps. In: Di Benedetto H, Baaj H,

Chailleux E, Tebaldi G, Sauzéat C, Mangiafico S (eds) Proceedings of the RILEM international symposium on bituminous materials—ISBM 2020, RILEM Bookseries vol 27. Springer, Cham, pp 215–221
37. EN 12697-8 (2020) Bituminous mixtures—test methods—part 8: determination of void characteristics of bituminous specimens. European Committee for Standardization, Brussels, Belgium
38. EN12697-23 (2017) Bituminous mixtures—test methods—part 23: determination of the indirect tensile strength of bituminous specimens. European Committee for Standardization, Brussels, Belgium
39. EN 12697-12 (2018) Bituminous mixtures. Test methods: determination of the water sensitivity of bituminous specimens. European Committee for Standardization, Brussels, Belgium
40. EN 12697-26 (2018) Bituminous mixtures—test methods—part 26: stiffness. European Committee for Standardization, Brussels, Belgium
41. EN 12697-25 (2016) Bituminous mixtures—test methods—part 25: cyclic compression test. European Committee for Standardization, Brussels, Belgium
42. EN 12697-31 (2019) Bituminous mixtures—test methods—part 31: specimen preparation by gyratory compactor. European Committee for Standardization, Brussels, Belgium
43. EN 12697-22 (2020) Bituminous mixtures—test methods—part 22: wheel tracking. European Committee for Standardization, Brussels, Belgium
44. EN 12697-33 (2019) Bituminous mixtures—test method—part 33: specimen prepared by roller compactor. European Committee for Standardization, Brussels, Belgium.
45. EN 13043 (2002) Aggregates for bituminous mixtures and surface treatments for roads, airfields and other trafficked areas. European Committee for Standardization, Brussels, Belgium

Chapter 5
Life Cycle Assessment of Asphalt Mixtures with WMR

Davide Lo Presti, Ana Jiménez del Barco Carrión, Gabriella Buttitta, Elizabeth Keijzer, Zhengyin Piao, Rita Kleizienė, Ben Moins, David Hernando, Emiliano Pasquini, Lily D. Poulikakos, Augusto Cannone Falchetto, and Di Wang

Abstract This study aimed at providing pavement specialists with international average values of greenhouse gas emissions (GHG) and energy consumption related to the manufacturing of asphalt mixtures (cradle-to-gate), with and without waste and secondary materials for roads (WMRs). In order to achieve this goal, TG leaders first referred to current standards to define a methodology for obtaining benchmark values according to ISO 21678:2020. Then, a data collection tool was designed and provided to a group of international partners to build a life cycle inventory for

D. Lo Presti (✉) · G. Buttitta
University of Palermo, Viale delle Scienze – Edificio 8, 90127 Palermo, Italy
e-mail: davide.lopresti@unipa.it

A. J. del Barco Carrión
LABIC-UGR. Laboratorio de Ingeniería de la Construcción Universidad de Granada ETS Caminos, Canales y Puertos Severo Ochoa S/N Campus de Fuentenueva, 18071 Granada, Spain

E. Keijzer
Netherlands Organisation for Applied Scientific Research (TNO), Oude Waalsdorperweg 63, 2597 AK Den Haag, The Netherlands

Z. Piao · L. D. Poulikakos
Empa, Swiss Federal Laboratories for Materials Science and Technology, Überlandstrasse 129, 8600 Dübendorf, Switzerland

R. Kleizienė
Road Research Institute of Vilnius TECH (Vilnius Gediminas Technical University), Saulėtekio Al. 11, 10223 Vilnius, Lithuania

B. Moins · D. Hernando
Energy and Materials in Infrastructure and Buildings (EMIB), University of Antwerp, Antwerp, Belgium

E. Pasquini
Department of Civil, Environmental and Architectural Engineering (ICEA), University of Padova, Via Marzolo, 9-35131 Padova, Italy

A. Cannone Falchetto · D. Wang
Department of Civil Engineering, Aalto University, Rakentajanaukio, 4-02150 Espoo, Finland

© The Author(s), under exclusive license to Springer Nature Switzerland AG 2023

A. Cannone Falchetto et al. (eds.), *Valorisation of Waste and Secondary Materials for Roads*, RILEM State-of-the-Art Reports 38,
https://doi.org/10.1007/978-3-031-33173-2_5

asphalt mixtures including data from materials up to plant manufacturing (A1–A3). Partners were asked to provide information for any type of available mixtures with and without WMRs. A total of 16 case studies were collected and subsequently energy consumption values were recorded, and several cradle-to-gate life cycle assessment exercises were performed to obtain GHG emissions values. This set of results was used to develop benchmark values for reference (conventional) asphalt mixtures, namely asphalt concretes (ACs) and stone matrix asphalts (SMAs). Finally, in order to assess how environmentally-friendly it is to incorporate WMRs into asphalt mixtures, the benchmark values of the reference asphalt concrete and stone matrix asphalt mixtures were compared with the GHG and energy consumptions of asphalt mixtures manufactured with selected WMRs. As a result: (1) SMA seems to have higher impact than ACs, (2) Asphalt mixtures engineered with selected WMRs seem to be always a more environmentally friendly option when compared to benchmark values of reference SMA. However, this does not always hold true when the comparison is made with conventional ACs.

Keywords Life Cycle Assessment · Benchmark · Asphalt mixture · Waste · Secondary materials

1 Introduction

Asphalt mixtures are the most used materials to pave roads around the world. During their whole life cycle, asphalt mixtures have associated a series of environmental impacts which must be controlled and reduced in order to meet current policies and sustainability strategies developed by governments.

The main environmental issues associated to the life cycle of asphalt mixtures are related to:

- the consumption of virgin (new) and non-renewable materials needed for producing new asphalt mixtures,
- the intense energy demand needed for their manufacturing at the high required temperatures,
- the greenhouse gases emissions associated to their transport, manufacturing and road construction,
- the consumption of more virgin and non-renewable materials for maintenance operations.

In order to solve these issues, different technologies are being studied and applied, such as the use of by-products and secondary materials to replace virgin materials, or the reduction of manufacturing temperatures to produce warm and cold asphalt mixtures. The application of these technologies must be assessed to check whether they are effectively able to reduce the mentioned environmental impacts.

In this regard, the evaluation of such impacts can be done by means of Life Cycle Assessment (LCA) techniques. LCA can be carried out with different approaches, methodologies and system boundaries. This exercise was carried out with a process-based attributional LCA limited to cradle-to-gate in order to gain information related to the impact of asphalt mixture up to manufacturing.

The following is the list of the participating organisations and the related acronyms:

1. University of Palermo (UNIPA)
2. University of Nottingham (UNOTT)
3. University of Granada (UGR)
4. Netherlands Organisation for Applied Scientific Research (TNO)
5. Road research institute of Vilnius TECH (Vilnius Gediminas technical university) (UVS)
6. University of Antwerp (UAW)
7. Swiss Federal Laboratories for Materials Science and Technology (EMPA).

1.1 Context

LCA is a well-accepted technique that was applied, improved and promoted for asphalt mixes over the last 20 years [1, 2]; during this time, framework documents were also developed and released [3]. Nevertheless, as there are currently no benchmarking schemes for evaluating the outputs or setting thresholds, explaining and evaluating LCA results remains challenging.

The ISO standard 21,678:2020 [4] describe (in Sect. 3.1) defines benchmarking as gathering, analyzing, and comparing performance data of equivalent buildings or other types of construction operations [4]. Initiatives to improve the environmental, social, and corporate governance (ESG) transparency of benchmark procedures and action to advance standards on low-carbon benchmarking in the European Union were anticipated in the action plan of the European Commission for financing sustainable growth (i.e., the E.U. taxonomy) [5]. In addition, the above-mentioned ISO standard [4] outlines the sources and categories of data that may be used for benchmarking as well as the guidelines and procedures for their declaration and communication for construction projects like buildings.

Literature on benchmark methods for LCA in the engineering and construction industry, mainly addressing buildings and structures, has received increasing attention, as demonstrated by the Athena Report devoted to LCA benchmarks on whole-building [6], the European SuPerBuildings Project [1], the Australian Materials and Buildings Products Life Cycle Inventory Database [7], the French "Construisons Ensemble HQE Performance" [8], and OneClickLCA [9]. Examples of large-scale efforts include the works of Simonen et al. [10] and Röck et al. [11], accounting for 1,191 and over 650 buildings in their evaluations, respectively.

At the national level, Germany implemented a rating system for Federal Buildings known as Bewertungssystem Nachhaltiges Bauen (BNB)—Assessment System for Sustainable Buildings [12]. Analogously, the Building Research Establishment

Environmental Assessment Method (BREEAM) supports the LCA of whole building [13]. Benchmarking activities are carried out by both systems [14].

Meanwhile, two case studies were found for benchmarking practices in the asphalt pavement literature. Saboori et al. [15] benchmarked the maintenance and rehabilitation actions at the end of the service life of asphalt pavements in California, while Butt et al. [16] benchmarked a runway reconstruction project at JKF airport in New York. However, the benchmarks in both cases are purely case specific and not intended to be sector-wide, as seen for buildings. The increased application of product category rules (PCRs) in the attempt to develop environmental product declarations (EPDs) is moving the mixture and paving industry toward a more harmonized process. For instance, within its PCRs, the National Asphalt Pavement Association (NAPA) acknowledged EPDs as appropriate for benchmarking efforts [17]. In the Sustainable Highway Construction Guidebook [18], it was also stated that EPDs could be used for benchmarking the impacts of a pavement or construction process. Although using EPDs leads to LCA outputs that are more harmonised, participation in the development of benchmarks is limited and especially when targeting a standardized and scientific approach that incorporates the current ISO standard 21,678:2020 [4]. Similar to the building industry, EPDs also do not take into account the issues related to data accessibility and availability in the paving sector.

1.2 Aim of the Study

In response to this need in the pavement sector and the lack of inclusion of the ISO 21678:2020 [4] standard requirements, TG4 was able to:

- Create a sector wide cradle-to-gate LCA benchmarking methodology for road pavement products that can be applied to reference asphalt mixtures and generate benchmarks.
- Collect case studies of asphalt mixtures with waste materials for roads (WMR), undertaking LCA and performing a comparative analysis with the benchmark values defined for reference asphalt mixtures.
- Provide some guidelines for the use of WMR asphalt mixtures in terms of reduction of their potential environmental impacts.

2 Methodology

The methodology of the TG was divided into two tasks, namely:

1. Benchmarking of reference asphalt mixtures
2. Environmental impact of asphalt mixtures with WMR

Figure 1 shows the structure of this TG. The benchmarks are first generated through the LCA of reference asphalt mixtures using the data provided by the participating

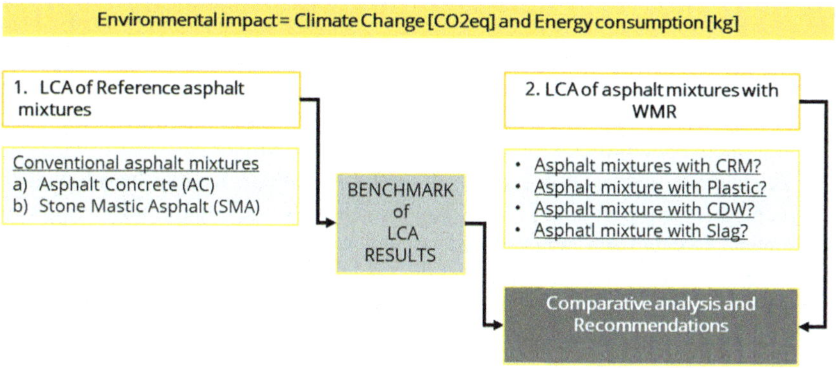

Fig. 1 TG4 methodology

organisations. Then, these values are compared with the results of the LCA of asphalt mixtures containing WMR. The two tasks of the methodology are described in detail in the following sections.

2.1 Benchmarking of Reference Asphalt Mixtures

In order to produce benchmarks that help assess the environmental impacts of asphalt mixtures by means of LCA, a literature review was conducted to identify available benchmarking methodologies and adapt them for road pavement products, particularly asphalt mixtures.

The keywords LCA, life cycle assessment, benchmark, and benchmarking were used in the Scopus database to identify publications of interest. A total of 605 documents from 1997 to 2020 were obtained, with 278 belonging to the engineering field. Projects and standards were searched using the Google Scholar database and the European standards database, respectively. To determine the methodological key variable and phases to perform the benchmarking for the LCA, every publication, project, and standard was thoroughly examined.

2.1.1 Key Methodological Variables

Five methodological key variables for benchmarking investigation were identified through the review of the chosen publications, and they are described below:

(1) *Study scope*: this variable provides the dimension of the analysis. This was measured by the amount of case studies taken into account, either through small-scale projects that only took into account a single case study or through large-scale research including several case studies.

(2) *Model typology*: the sample selected for the analysis may differ across three categories: (a) Samples belonging to the first type are external and rely on real case studies; (b) internal samples are derived from standards on construction and are used for the pre-construction phase [19]; (c) the last type of samples are obtained by combining the external and internal samples.
(3) *Benchmark approach*: top-down or bottom-up strategies can be adopted for benchmarking purposes [1, 20]. Top-down refers to the definition of emission inventory data from industry-wide statistics or goals established by legislation (such as setting standards to meet the criteria of the Paris Climate Agreement announced in UNFCCC in 2015). Bottom-up models of present assets are developed from actual experience or theoretical models (e.g. Modelling the defining characteristics of the model typology) [6].
(4) *Database selection*: it represents the data resources for benchmarking. It is crucial for developing a standard concerning the quality or validity of data; it typically consists of commercial or open-source data sources, original databanks, or literature studies.
(5) *Type of benchmarks*: it consists of the numerical evaluations provided to clients or end users. While most of the analysis only offered a single output, this can increase to five depending on the studies. Three essential types of benchmark values can be identified based on the ISO standard 21,678:2020 [4]: limit (maximum undesired value), reference (current practice value), and target (ideal practice value).

The definition of the five variables listed above enabled the identification of the most common methods presented in the literature, which indicated to be those based on large-scale analysis relying on built (e.g., external) models. The LCA is often conducted through a bottom-up methodology (comparable to a process-based LCA), using data obtained from third-party sources. A few extensive studies could also develop a dedicated database for the environmental impact tailored to the particular project. Considering that the base number of benchmarks indicated by the ISO standard 21,678 [4] is three, only 22% of the work in literature met this recommendation, while 39% of the documents presented a single benchmark.

2.1.2 General Steps Required for Benchmarking of LCA Results

The literature review carried out revealed that three steps are needed for creating benchmarks. The three key steps are (1) data collection, (2) LCA of case studies, and (3) benchmark generation. Figure 2 shows a summary of these steps and which methodological variables have to be considered in each step. Figure 2 also includes the different options that can be adopted for each methodological variable and whether they were found to be the most recommended, acceptable or not recommended.

1. **Data Collection**: this first step focuses on the creation of a case study database for benchmarking. Before the database can be created, the goal and scope of the study must be defined. Upon their definition, case studies can start to be

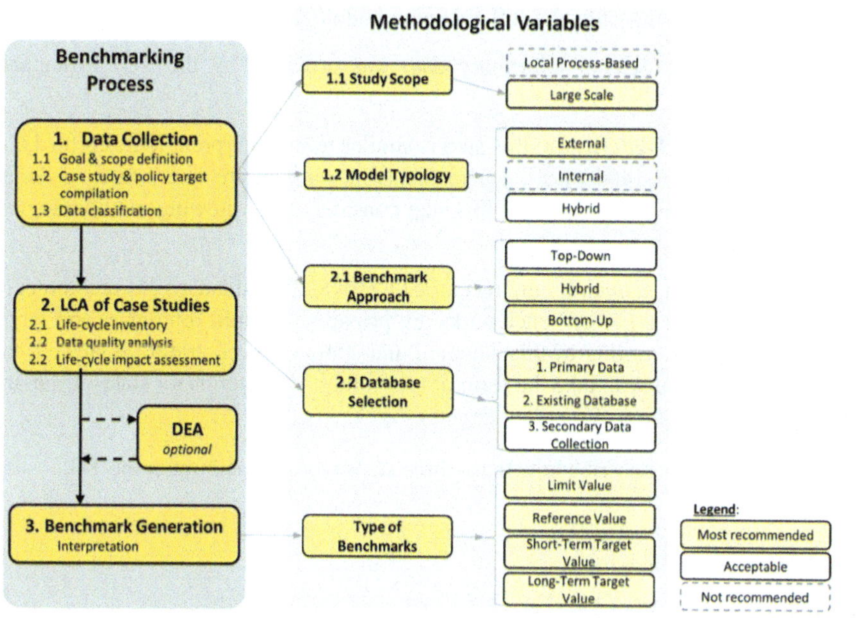

Fig. 2 General steps required for LCA benchmarking

compiled. If target values are part of the scope of the study, relevant policy targets are collected too. After collecting case studies, they are classified according to the key benchmark parameters of interest defined in the scope.

As part of this step, the following key methodological variables were considered:

a. *Study Scope*: a large-scale study is recommended for a sector-wide initiative. Under this scope, various case studies were considered, varying in life cycle inventory and region.
b. *Model Typology*: an external model typology (i.e., post-construction) was selected, creating a more realistic case study database.
c. *Benchmark Approach*: a bottom-up approach was selected to calculate the limit, reference and target value benchmarks. Despite involving further complexity and time requirements, it is more precise and process specific.

2. **LCA of Case Studies**: this step calculates the environmental impact of the case studies. Firstly, a life-cycle inventory (LCI) is compiled for all products and processes in the case study database (either through data collection or third-party data). Then, the impact assessment takes place to quantify the impacts of each case study. The benchmarking database should contain material quantities as well as environmental burdens, to ensure the database's updateability when the product and process impacts are updated [21].

As part of this step, the following methodological variable was considered:

d. *Database Selection*: primary data was provided by the TG participants [14, 22].

3. **Benchmark Generation**: this step is similar to the interpretation step for LCA from the ISO 14040 [23], where the LCA outputs are interpreted to define the benchmark values. Thus, this step considers the following methodological variable:

   e. *Type of benchmarks*: the recent ISO 21678:2020 standard [4] recommends that three types of benchmarks are provided: a limit, reference, and target value. According to this standard and some of the literature reviewed, the target value can take the form of two values: short-term (for state-of-the-art) and long-term (for policy targets).

The next sections describe how these three steps were carried out in the TG.

2.1.3 Data Collection

There exist two main obstacles for data collection when performing an LCA: 1) lack of knowledge of the type of data needed to perform the LCA; 2) lack of availability of these data. Therefore, a series of webinars were organised to start training the TG members on how to perform LCA and what type of data is needed. Then, the "**RILEM TC WMR TG4 LCA Data Collection Tool**" was developed and distributed to all TG members to facilitate data collection, both for the TG and for any future LCA of asphalt mixtures that they may undertake.

The RILEM TC WMR TG4 LCA Data Collection Tool is a compressed file containing eight files. The first file is a PDF providing the context of the TG and describes the main features of the tool. The rest of the files are Excel spreadsheets in which the necessary data can be collected for different case studies. The seven Excel files are organised in the same way, with the following tabs:

- Cover
- Instructions
- General Info
- Raw Material Supply (A1)
- Transport to Plant (A2)
- Manufacturing (Plant) (A3)
- Previous LCA—Cradle to gate

Each tab of the "RILEM TC WMR TG4 LCA Data Collection Tool" is described in the following sections.

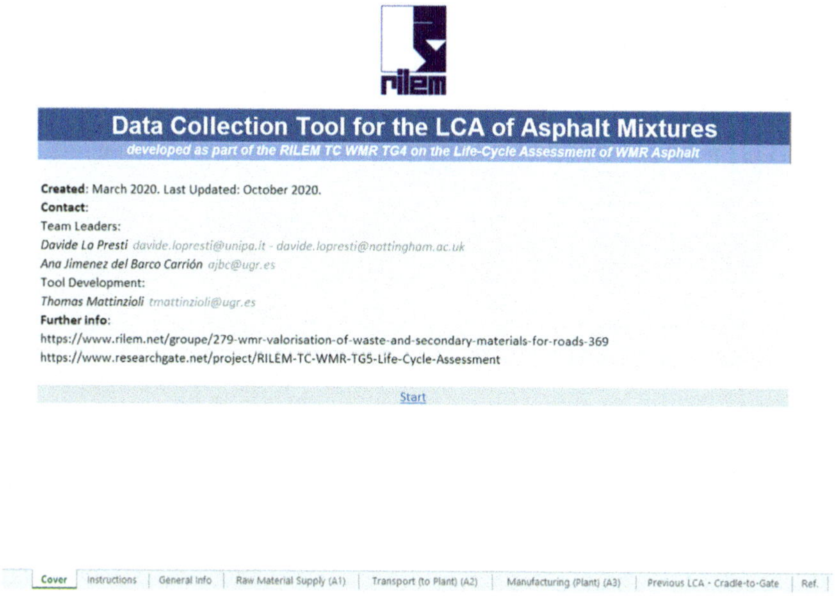

Fig. 3 Cover tab of the "RILEM TC WMR TG4 LCA data collection tool"

Cover

Figure 3 shows a screenshot of the Cover tab within the "RILEM TC WMR TG4 LCA Data Collection Tool". This tab provides the user with information about the tool version, the contact information of the creators, and links with further details about the TC. It contains a "Start" button to click and move to the Instructions tab.

Instructions

This tab (Fig. 4) provides the information needed to use the tool successfully. It also contains a link to the PDF file with the general overview of the TG and the tool and a "Next" button to click and move to the General Info tab.

General Info

The General Info tab gives information about the participant using the tool and the case study for which the data is being provided. This information includes the type of asphalt mixture, the production year, location, and energy sources used (Fig. 5). If known, some extra details on the road for which the asphalt mixture is manufactured can be provided. It also contains "Back" and "Next" buttons to click and move to the Instructions or Raw Material Supply tabs, respectively.

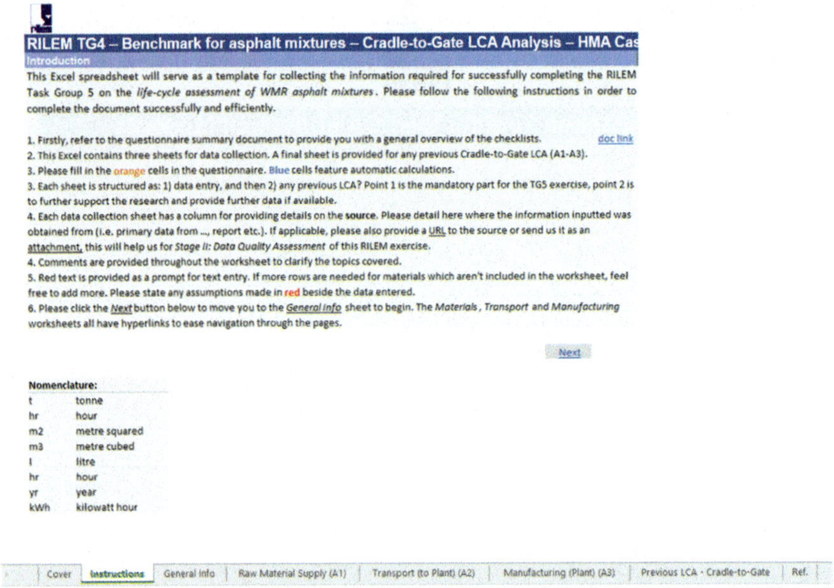
RILEM TG4 – Benchmark for asphalt mixtures – Cradle-to-Gate LCA Analysis – HMA Cas

Introduction

This Excel spreadsheet will serve as a template for collecting the information required for successfully completing the RILEM Task Group 5 on the *life-cycle assessment of WMR asphalt mixtures*. Please follow the following instructions in order to complete the document successfully and efficiently.

1. Firstly, refer to the questionnaire summary document to provide you with a general overview of the checklists. doc link
2. This Excel contains three sheets for data collection. A final sheet is provided for any previous Cradle-to-Gate LCA (A1-A3).
3. Please fill in the orange cells in the questionnaire. Blue cells feature automatic calculations.
3. Each sheet is structured as: 1) data entry, and then 2) any previous LCA? Point 1 is the mandatory part for the TG5 exercise, point 2 is to further support the research and provide further data if available.
4. Each data collection sheet has a column for providing details on the **source**. Please detail here where the information inputted was obtained from (i.e. primary data from ..., report etc.). If applicable, please also provide a URL to the source or send us it as an attachment, this will help us for *Stage II: Data Quality Assessment* of this RILEM exercise.
4. Comments are provided throughout the worksheet to clarify the topics covered.
5. Red text is provided as a prompt for text entry. If more rows are needed for materials which aren't included in the worksheet, feel free to add more. Please state any assumptions made in red beside the data entered.
6. Please click the *Next* button below to move you to the *General Info* sheet to begin. The *Materials*, *Transport* and *Manufacturing* worksheets all have hyperlinks to ease navigation through the pages.

Next

Nomenclature:

t	tonne
hr	hour
m2	metre squared
m3	metre cubed
l	litre
hr	hour
yr	year
kWh	kilowatt hour

Cover | Instructions | General Info | Raw Material Supply (A1) | Transport (to Plant) (A2) | Manufacturing (Plant) (A3) | Previous LCA - Cradle-to-Gate | Ref.

Fig. 4 Instructions tab of the "RILEM TC WMR TG4 LCA data collection tool"

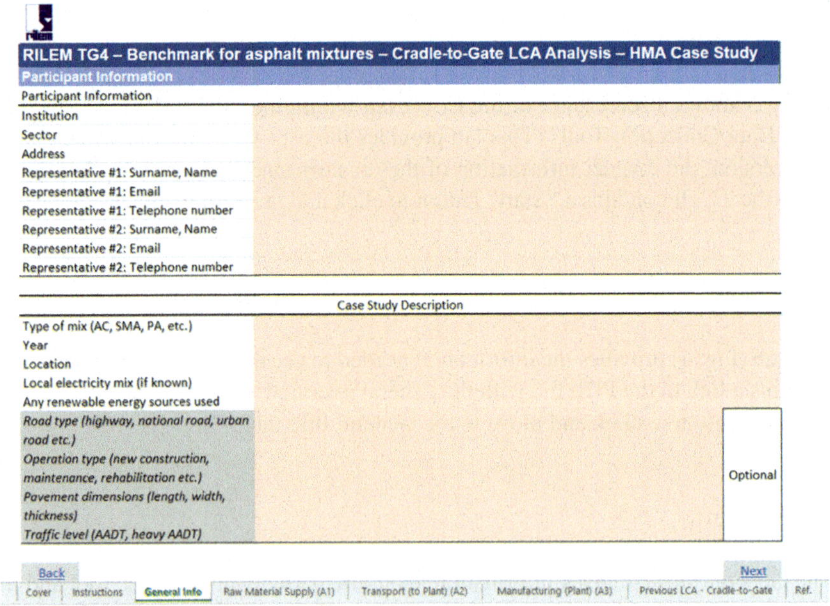
RILEM TG4 – Benchmark for asphalt mixtures – Cradle-to-Gate LCA Analysis – HMA Case Study

Participant Information

Participant Information	
Institution	
Sector	
Address	
Representative #1: Surname, Name	
Representative #1: Email	
Representative #1: Telephone number	
Representative #2: Surname, Name	
Representative #2: Email	
Representative #2: Telephone number	

Case Study Description		
Type of mix (AC, SMA, PA, etc.)		
Year		
Location		
Local electricity mix (if known)		
Any renewable energy sources used		
Road type (highway, national road, urban road etc.)		Optional
Operation type (new construction, maintenance, rehabilitation etc.)		Optional
Pavement dimensions (length, width, thickness)		Optional
Traffic level (AADT, heavy AADT)		Optional

Back Next

Cover | Instructions | General Info | Raw Material Supply (A1) | Transport (to Plant) (A2) | Manufacturing (Plant) (A3) | Previous LCA - Cradle-to-Gate | Ref.

Fig. 5 General Info tab of the "RILEM TC WMR TG4 LCA data collection tool"

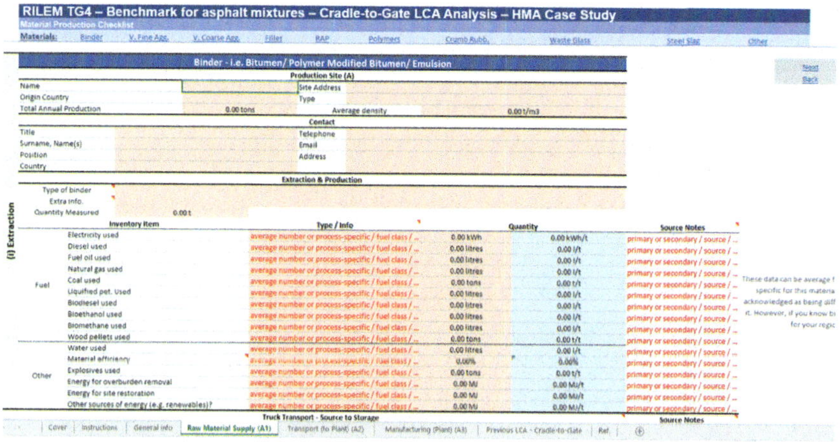

Fig. 6 Raw material supply (A1) tab of the "RILEM TC WMR TG4 LCA data collection tool"

Raw Material Supply (A1)

This tab (Fig. 6) is used to collect all the necessary data about each raw material used for the case study (i.e., binder, aggregates, filler, polymers, etc.). The input data includes the production site of the material, contact, extraction, production (energy consumption), transport to storage, storage, and whether any previous LCA results are available. This tab allows easy navigation by clicking on the specific material at the top of the tab. It also contains "Back" and "Next" buttons to click and move to the General Info or Transport tabs, respectively.

Transport (to Plant) (A2)

This tab, depicted in Fig. 7, reports the data related to transport distances between the quarry/storage and the asphalt plant where the mixture will be produced. The information is related to origin, destination, type of transport (truck and/or inland ship), and the number of kilometres travelled. It also contains "Back" and "Next" buttons to click and move to the Raw Material Supply or Manufacturing tabs, respectively.

Manufacturing (A3)

Figure 8 illustrates the tab used to collect all the information about the mix design and production of the mix. Hence, the data required is linked to the amount and type of materials used (i.e., virgin aggregates, binder, filler, additives, etc.) and the type and quantity of energy needed (e.g., diesel, natural gas, etc.).

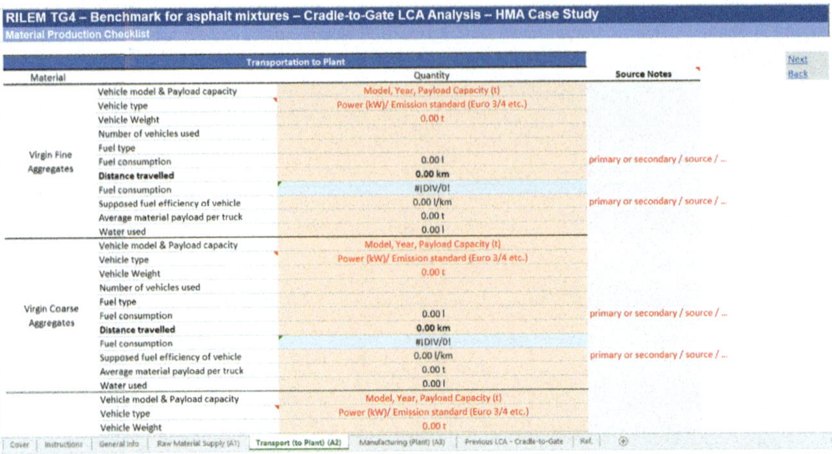
RILEM TG4 – Benchmark for asphalt mixtures – Cradle-to-Gate LCA Analysis – HMA Case Study
Material Production Checklist

Transportation to Plant

Material		Quantity	Source Notes
Virgin Fine Aggregates	Vehicle model & Payload capacity	Model, Year, Payload Capacity (t)	
	Vehicle type	Power (kW)/ Emission standard (Euro 3/4 etc.)	
	Vehicle Weight	0.00 t	
	Number of vehicles used		
	Fuel type		
	Fuel consumption	0.00 l	primary or secondary / source / ...
	Distance travelled	**0.00 km**	
	Fuel consumption	#DIV/0!	
	Supposed fuel efficiency of vehicle	0.00 l/km	primary or secondary / source / ...
	Average material payload per truck	0.00 t	
	Water used	0.00 l	
Virgin Coarse Aggregates	Vehicle model & Payload capacity	Model, Year, Payload Capacity (t)	
	Vehicle type	Power (kW)/ Emission standard (Euro 3/4 etc.)	
	Vehicle Weight	0.00 t	
	Number of vehicles used		
	Fuel type		
	Fuel consumption	0.00 l	primary or secondary / source / ...
	Distance travelled	**0.00 km**	
	Fuel consumption	#DIV/0!	
	Supposed fuel efficiency of vehicle	0.00 l/km	primary or secondary / source / ...
	Average material payload per truck	0.00 t	
	Water used	0.00 l	
	Vehicle model & Payload capacity	Model, Year, Payload Capacity (t)	
	Vehicle type	Power (kW)/ Emission standard (Euro 3/4 etc.)	
	Vehicle Weight	0.00 t	

Next
Back

Cover | Instructions | General Info | Raw Material Supply (A1) | Transport (to Plant) (A2) | Manufacturing (Plant) (A3) | Previous LCA - Cradle-to-Gate | Ref.

Fig. 7 Extract of transport to plant (A2) tab of the "RILEM TC WMR TG4 LCA data collection tool"

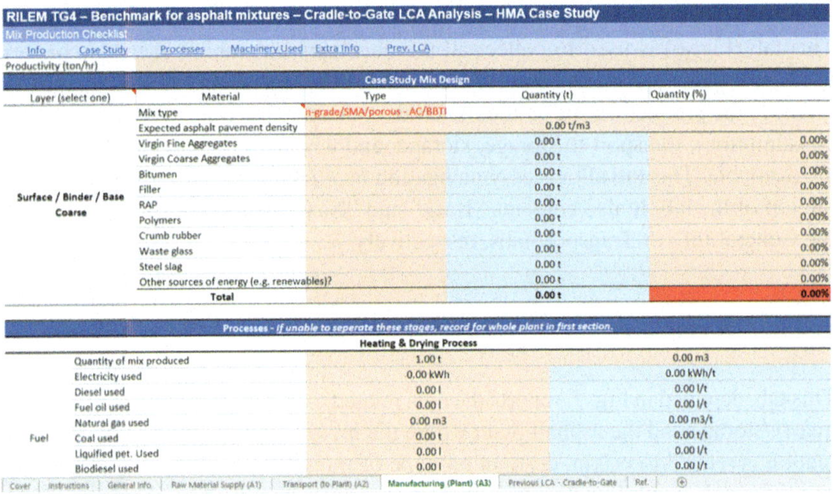
RILEM TG4 – Benchmark for asphalt mixtures – Cradle-to-Gate LCA Analysis – HMA Case Study
Mix Production Checklist
Info Case Study Processes Machinery Used Extra Info Prev. LCA
Productivity (ton/hr)

Case Study Mix Design

Layer (select one)	Material	Type	Quantity (t)	Quantity (%)
Surface / Binder / Base Coarse	Mix type	n-grade/SMA/porous - AC/BBTI		
	Expected asphalt pavement density		0.00 t/m3	
	Virgin Fine Aggregates		0.00 t	0.00%
	Virgin Coarse Aggregates		0.00 t	0.00%
	Bitumen		0.00 t	0.00%
	Filler		0.00 t	0.00%
	RAP		0.00 t	0.00%
	Polymers		0.00 t	0.00%
	Crumb rubber		0.00 t	0.00%
	Waste glass		0.00 t	0.00%
	Steel slag		0.00 t	0.00%
	Other sources of energy (e.g. renewables)?		0.00 t	0.00%
	Total		**0.00 t**	**0.00%**

Processes - *If unable to seperate these stages, record for whole plant in first section.*

Heating & Drying Process

Fuel	Quantity of mix produced	1.00 t	0.00 m3
	Electricity used	0.00 kWh	0.00 kWh/t
	Diesel used	0.00 l	0.00 l/t
	Fuel oil used	0.00 l	0.00 l/t
	Natural gas used	0.00 m3	0.00 m3/t
	Coal used	0.00 t	0.00 t/t
	Liquified pet. Used	0.00 l	0.00 l/t
	Biodiesel used	0.00 l	0.00 l/t

Cover | Instructions | General Info | Raw Material Supply (A1) | Transport (to Plant) (A2) | Manufacturing (Plant) (A3) | Previous LCA - Cradle-to-Gate | Ref.

Fig. 8 Extract of manufacturing (A3) tab of the "RILEM TC WMR TG4 LCA data collection tool"

Previous LCA—Cradle-to-Gate

This last tab (Fig. 9) aims to collect previous LCA results if they were available for the case study, and if that is the case, compare them with the results obtained in this analysis.

RILEM TG4 – Benchmark for asphalt mixtures – Cradle-to-Gate LCA Analysis – HMA Case Study

Cradle-To-Gate LCA Results

Details of Cradle-to-gate Case Study

Please provide as much detail as possible:

(A1) Mixture type and breakdown of materials used (wt%)

(A2) Transportation fleet, average fuel consumption (loaded and unloaded)

(A3) Plant type, annual production, fuel type, manufacturing temperature, productivity (ton/hr), fuel consumption (fuel/hr - fuel/ton)

CRADLE-TO-GATE LCA Results

Impact	Quantity	Unit	Source	Impact Assessment Methodology Used	Indicator Description Reference
Energy	0.00	MJ/t			EN 15804:2012, Table 4 / ReCiPe 2016
Climate Change	0.00	kg CO2 eq./t			EN 15804:2012, Table 3 / ReCiPe, 2016
Total Material Use	0.00	kg/t			EN 15804:2012, Table 8 / ReCiPe, 2016
Virgin Material Use	0.00	kg/t			EN 15804:2012, Table 8 / ReCiPe, 2016
Secondary/ Recycled	0.00	kg/t			EN 15804:2012, Table 4 / ReCiPe, 2016
Photochemical Oxidation	0.00	kg ethylene eq. (/kg NMVOC)/t			EN 15804:2012, Table 3 / ReCiPe, 2016
Acidification	0.00	kg SO2 eq./t			EN 15804:2012, Table 3 / ReCiPe, 2016
Eutrophication	0.00	kg PO4 eq. (/kg N eq.)/t			EN 15804:2012, Table 3 / ReCiPe, 2016
Depletion of Abiotic Resources: elements/ fossil fuels	0.00	kg Sb eq. (/MJ, kg antimony equivalent/ kg minerals/ MJ			EN 15804:2012, Table 3

Cover | Instructions | General info | Raw Material Supply (A1) | Transport (to Plant) (A2) | Manufacturing (Plant) (A3) | Previous LCA · Cradle to Gate | Ref.

Fig. 9 Extract of existing cradle-to-gate results tab of the "RILEM TC WMR TG4 LCA data collection tool"

Data Collected

Using the "RILEM TC WMR TG4 LCA Data Collection Tool", data was collected thanks to the contribution of the participants of the TG, who provided primary data for fifteen case studies of conventional asphalt mixtures. Examples are shown in Figs. 10 and 11 as examples.

2.1.4 LCA of Reference Asphalt Mixtures Case Studies

Three life cycle stages were analysed to shed light on the environmental impacts of asphalt mixtures. These are: Raw Material Supply (A1), Transport to Plant (A2), and Manufacturing (A3) (Fig. 12). Hence, the analysis performed is a so-called "cradle-to-gate": from the extraction of raw materials ("cradle") to the product ready to be used and distributed ("gate" of the asphalt plant). These life cycle stages correspond to the life cycle of asphalt mixtures as road pavement components before a service life or durability is assigned to them. Consequently, the performed analysis focuses on the materials, their extraction, and design. In addition, previous research has shown that A1-A3 life stages can contribute between 69% and 82% of the total impact of the life cycle of pavement materials [24].

As shown in Fig. 13, two reference asphalt mixtures were considered: asphalt concrete (AC) and stone mastic asphalt (SMA).

LCA allows calculating the environmental burdens due to the selected life cycle stages. The LCA was performed compounding the ISO standards 14,040 [23] and 14,044 [25] and following the four prescribed steps: 1. Goal and Scope Definition; 2. Life Cycle Inventory; 3. Life Cycle Impact Assessment; 4. Interpretation of results.

Institution	Country	Year	Binder: Origin	Destination	Truck		Inland Ship	Nautical Ship
					km	Emissions Standard	km	km
University of Antwerp	Belgium	2019	Vitol		61.00 km	EURO 5		
University of Antwerp	Belgium	2019	Vitol		61.00 km	EURO 5		

Fine Aggregates - Sand: Origin	Destination	Distance Travelled	Truck		Inland Ship	Nautical Ship
			Sec. Distance	Emissions	km	km
Sagrex		128.00 km		EURO 5		
Sagrex		128.00 km		EURO 5		

Coarse Aggregates - Gravel: Origin	Destination	Truck		Inland Ship	Nautical Ship
		Sec. Distance	Emissions Standard	km	km
		128.00 km	EURO 5		
		128.00 km	EURO 5		

Filler: Origin	Destination	Truck	
		km	Emissions Standard
CBM		124.00 km	EURO 5
CBM		124.00 km	EURO 5

Fig. 10 Data collection about transport to plant (A2) for two case studies: UAW_AC10 (top row) and UAW_AC20 (bottom row)

Goal and Scope Definition

This part of the study aims to calculate the environmental impacts of different mixtures during their product stage (A1–A3) so benchmarks can be later defined.

The intended applications of the study are:

- The understanding of environmental burdens due to the production of asphalt mixtures;
- The support of a more sustainable decision-making process in the asphalt industry;
- The understanding of benchmarking values.

Institution	Country	Year	Asphalt Plant Information: Address	Type	Model	Year	Mix Type	Course	Density	Binder Type	Binder %
University of Antwerp	Belgium	2019	Belasco	Batch	Ammann	1994	AC 10	Surface/ Wearing	2.37 t/m3	PEN 50/70	5.93%
University of Antwerp	Belgium	2020	Gaurain-Ramecroix	Batch	Ammann	1994	AC 20	Base	2.35 t/m3	PEN 50/70	4.82%

Case Study Mix Design: Filler Type	Filler %	Fine Agg. *Crushed* Type	Fine Agg. *Crushed* %	Fine Agg. *Natural* Type	Fine Agg. *Natural* %	Coarse Agg. *Crushed* Type 1	%	Type 2	%
0	6.08%	Sandstone 0/2, 2/6.3	32.2%	0		Sandstone 6.3/10	55.75%	0	
0	5.20%	Limestone	54.3%	0		Limestone	36.33%	0	

Production: Temperature	Productivity	Electricity	Natural Gas	Fuel Oil	Diesel	Primary Data?
178 °C	150.0 t/hr	6.12 kWh/t	0.00 m3/t	8.50 l/t		Yes
185 °C	320.0 t/hr	6.12 kWh/t	8.66 m3/t	0.00 l/t		Yes

Fig. 11 Data collection about manufacturing (A3) for two case studies: UAW_AC10 (top row), UAW_AC20 (bottom row)

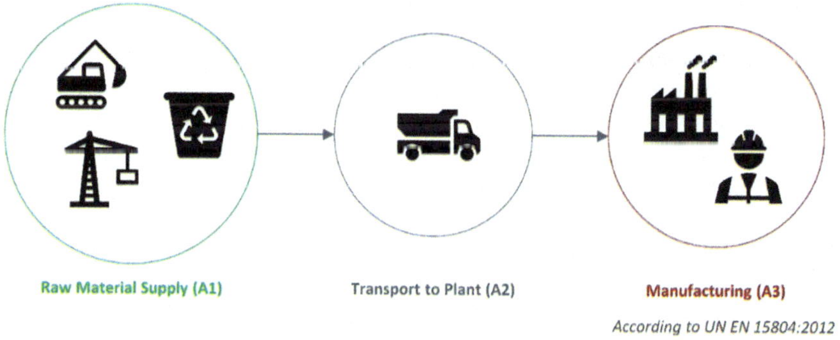

Fig. 12 Life cycle stages of asphalt mixtures as road pavement materials

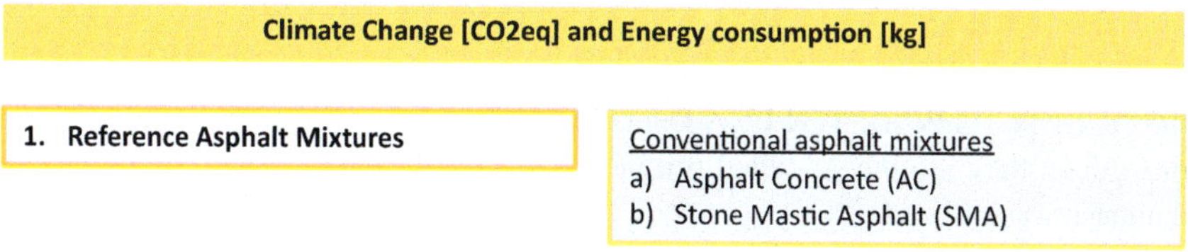

Fig. 13 Structure of the benchmarking phase 2: LCA of case studies

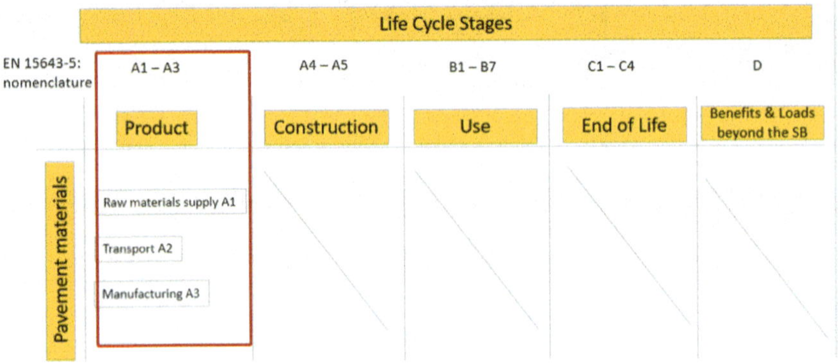

Fig. 14 System boundaries and life cycle stages (adapted from Pavement LCM [26])

The provided mix designs are for mixtures produced in plants in different European countries, and TG partners have provided all the information.

As required by the standards, the following declared unit was chosen: 1t of asphalt mixture to be produced. The system boundaries are A1-A3 stages, as shown in Fig. 14 and detailed below:

- A1: Raw materials acquisition (i.e., mining, manufacturing)
- A2: Transport (i.e., trucks and diesel necessary for the transport)
- A3: Production (i.e., diesel/electricity used)

Concerning the cut-off rules, none was applied to consider the complete material impact. The selected impact categories are related to energy resources and climate change. The first one, measured in kg, represents the quantity of nonrenewable and renewable energy resources needed for all the processes (e.g., crude oil, peat, lignite), as reported in Fig. 15. The second one is calculated through the EF 2.0 methodology developed by the European Commission in 2018. The conversion factor from kg to mJ is 46.

Concerning the data, they must comply with certain IS014040:2006 requirements [23]. This is detailed in Life Cycle Inventory (LCI) section.

Life Cycle Inventory

This step in the LCA is strictly connected to the data collection, performed using the "RILEM TC WMR TG4 LCA Data Collection Tool". Partners directly provided most data: they are the so-called primary data. When it was not available, GaBi ts database was utilized.

In total, fourteen case studies were analyzed. Table 1 shows the mix design, while Table 2 presents the transport and energy used in each case study.

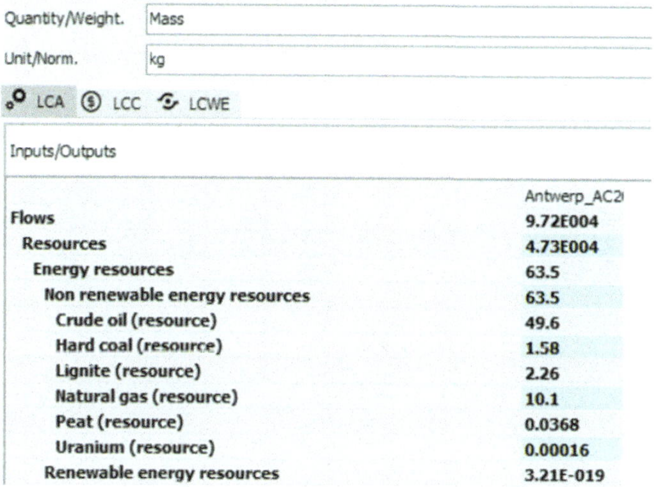

Fig. 15 Screenshot from GaBi concerning the energy resources

Three scenarios were analysed in terms of transport distances to compare the different case studies and listed as follows:

1. Real transport distances
2. Transport distances of 50 km for trucks and 600 km for ships;
3. Transport distances of 150 km for trucks and 1000 km for ships.

This allowed performing a sensitivity analysis on the contribution of transport distances. In this research, transport distances are an interesting parameter to look at since they are very different for each case study, and in some of them, inland shipping was considered.

Life Cycle Impact Assessment

For the purpose of this investigation, two life cycle impact indicators were calculated: Climate Change and Energy Used. The calculation was performed using the EF2.0 methodology, as European standards suggest, available in GaBi ts software. No optional step (normalization or weighting) was carried out. The selection of these indicators results from previous research showing their importance in the environmental impact of asphalt mixtures [23, 27].

In this specific research, the calculations were performed modelling the life cycle stages of each case study using GaBi ts (Fig. 16). The mentioned software allows to input of all the previously collected information and assesses the environmental burdens according to a specific calculation methodology. The methodology selected is the EF 2.0, as the current European Standards suggest.

Table 1 Mix design of each case study

Case study	Fine aggregates		Coarse aggregates		Filler		Binder	
	Type	Percentage in the mix	Type	Percentage in the mix	Type	Percentage in the mix	Type	Percentage in the mix
TNO_Open Stone	Nat.+crus. sand	12.89 44.01	Crushed stone	37.56	Limestone	2.62	70/100	2.92
UNIPA_ AC32TS	Crushed sand	29.5	Crushed stone	60.0	Limestone	7.0		3.5
UVS_ AC22PS	Dolomite Sand (0/2)	20 20	Gravel chip Granite chip	25.74 10.48	Limestone	0.95	50/70	3.80
UAW_ AC20	Limestone	53.95	Sandstone 6.3/10	36.09	Limestone	5.17	50/70	4.79
UNOTT_ UK	Crushed sand	35	Crushed stone	52.4	Limestone	2.90	70/100	5.10
UNIPA_ AC12	Crushed sand	40	Crushed stone	50.5	Limestone	4.0	45/80–70	5.5
UVS_AC11	Dolomite Sand (0/5)	19.3 15.1	Dolomite	33.90	Limestone	7	70/100	5.80
TNO_AC SURF	Nat.+crus. sand	28.35 9.35	Crushed stone	39.63 11.79	Limestone	4.98	40/60	5.89
UAW_ AC10	Sandstone 0/2, 2/ 6.3	32.24	Stone 6.3/10	55.75	Limestone	6.08	50/70	5.93
UNIPA_ ITALY	Crushed sand	32.9	Crushed stone	5.20	Limestone	7	70/100	6.10
TNO_WB AC	Sand	37.60	Crushed stone	51.22	Limestone	4.98	70/100	6.20

(continued)

Table 1 (continued)

Case study	Fine aggregates		Coarse aggregates		Filler		Binder	
	Type	Percentage in the mix	Type	Percentage in the mix	Type	Percentage in the mix	Type	Percentage in the mix
TNO_SMA GR	Nat.+crus. sand	5.46 4.85	Crushed stone	75.89	Limestone	7.16	70/100	7.30
TNO_SMA 5	Nat.+crus. sand	7.92 5.87	Crushed stone	68.94	Limestone	10.1	70/100	7.26
TNO_SMA 8-11	Nat.+crus. sand	8.30 8.30	Crushed stone	74.78	Limestone	1.33	70/100	7.30

Table 2 Transport and energy used of each case study

Case study	Transport of fine aggregates		Transport of coarse aggregates		Transport of filler		Transport of binder		Energy used	
	Type	Distance (km)	Type	Distance (km)	Type	Distance (km)	Type	Distance (km)	Type	Amount
TNO_Open Stone[1]	Euro V	150	Euro V	25 986 150	Euro V	136	Euro V	89	Electricity Natural gas Diesel	4.93 kWh/t 8.82 m^3/t 0.12 l/t
UNIPA_ AC32TS	Euro V	28	Euro V	28	Euro V	28	Euro V	163	Oil Electricity	10 kg/t 3.33 kWh/t
UVS_AC22PS	Euro V	39	Euro V	75	Euro V	104	Euro V	460	Electricity	2.18 KWh/t
UAW_AC20	Euro V	128	Euro V	128	Euro V	124	Euro V	61	Electricity Natural gas	6.12 kWh/t 8.66 m^3/t
UNOTT_UK	Euro V	70	Euro V	70	Euro V	100	Euro V	160	Electricity Diesel	4.80 kWh/t 1.60 l/t
UNIPA_AC12	Euro V	65	Euro V	235	Euro V	150	Euro V	125	Electricity Natural gas Diesel	6.75 kWh/t 10 m^3/t 10 l/t
UVS_AC11[1]	Euro V	80	Euro V	39 24	Euro V	60	Euro V	50	Electricity Natural gas	2.23 kWh/t 13.26 m^3/t
TNO_AC SURF[1]	Euro V	25 986 150	Euro V	25 986 150	Euro V	136	Euro V	89	Electricity Natural gas Diesel	6.61 kWh/t 8.81 m^3/t 0.12 l/t
UAW_AC10	Euro V	128	Euro V	128	Euro V	124	Euro V	61	Electricity	6.12 kWh/t
UNIPA_ITALY	Euro V	23	Euro V	23	Euro V	23	Euro V	268	Electricity Diesel	4.80 kWh/t 1.60 l/t

(continued)

Table 2 (continued)

Case study	Transport of fine aggregates		Transport of coarse aggregates		Transport of filler		Transport of binder		Energy used	
	Type	Distance (km)	Type	Distance (km)	Type	Distance (km)	Type	Distance (km)	Type	Amount
TNO_WB AC[1]	Euro V	150	Euro V	25 986 150	Euro V	136	Euro V	89	Electricity Natural gas Diesel	6.79 kWh/t 8.75 m^3/t 0.12 l/t
TNO_SMA GR[1]	Euro V	25 986 150	Euro V	25 986 150	Euro V	136	Euro V	89	Electricity Natural gas Diesel	7.06 kWh/t 7.37 m^3/t 0.12 l/t
TNO_SMA 5[1]	Euro V	25 986 150	Euro V	25 986 150	Euro V	136	Euro V	89	Electricity Natural gas Diesel	7.42kWh/t 7.37m^3/t 0.12 l/t
TNO_SMA 8 = 11[1]	Euro V	25 986 150	Euro V	25 986 150	Euro V	136	Euro V	89	Electricity Natural gas Diesel	7.08 kWh/t 8.04 m^3/t 0.12 l/t

[1] Aggregate shipped to TNO and UVS are sourced from different places

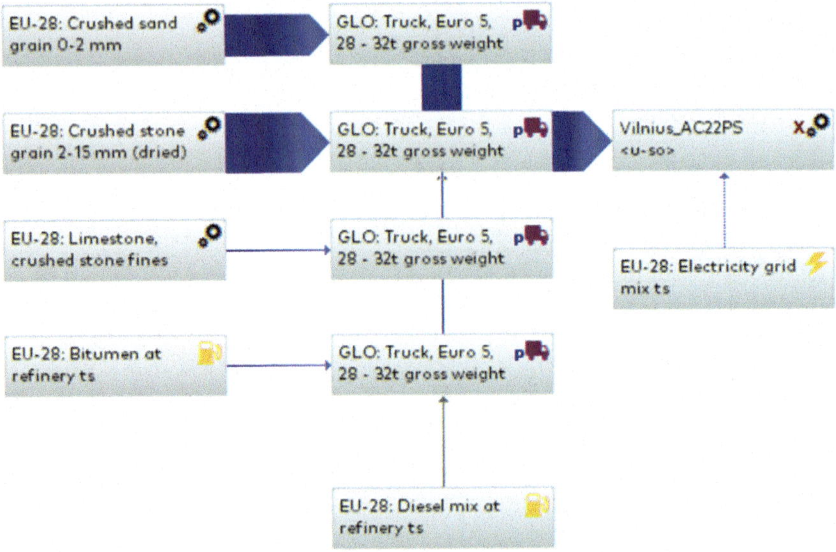

Fig. 16 Life cycle modelling produced in GaBi ts software for the reference mixtures

Interpretation of Results

In this part, the results are discussed and summarised, in accordance with the goal and scope of this LCA. The following analysis was carried out:

- Hotspot analysis

This analysis is performed to identify hotspots in the life cycle of a product. According to a JRC Report [28], hotspots in LCA are those life cycle stages or processes that contribute in an important way to the whole impact. In particular, a hotspot in LCA is defined as the life cycle stage or process that has an influence of more than 50%, while it is defined as relevant if its contribution is more than 80%.

The contribution of each life stage (A1, A2, A3) was calculated for all the mixtures with the actual and the assumed distances.

- Sensitivity analysis

The assumption of different transport distances allowed to understand the impact of transport in terms of energy consumption and climate change.

2.1.5 Benchmark Generation

As mentioned in Sect. 2.1.2, the recent ISO 21678:2020 [4] standard recommends that three types of benchmarks are provided: a limit, a reference, and a target value. According to this standard, and in some of the literature reviewed, the target value can take the form of two values: short-term (for state-of-the-art) and long-term (for policy targets).

According to the ISO 21678:2020 [4], the reference values can be represented by either the mean, median, or modal values, by specific percentile values, or by technical and/or economic optimum or feasibility [4]. In this investigation, the reference value was calculated using the mean of the results for the impact category indicators assuming reference asphalt mixtures with 50 km as transport distances. These reference values (RF) are determined as shown in Eqs. (1) and (2) for Climate Change and Energy Use, respectively.

$$RF_{CG} = \frac{\sum_{n}^{i=1} CC_i}{n} \quad (1)$$

$$RF_{EU} = \frac{\sum_{n}^{i=1} EU_i}{n} \quad (2)$$

where,

RF_{CG} reference value for Climate Change,
RF_{EU} reference value for Energy Use,
CC_i Climate Change impact indicator of case study *i*,
EU_i Energy Use impact indicator of case study *i*,
n total number of case studies for reference asphalt mixtures.

After creating the reference benchmark value, the generation of limit and target values is required. According to ISO 21678:2020 [4], limit values can be obtained from statistical analysis, surveys, theoretical calculations, legal requirements, and standards (i.e., from both internal and external models). Target values may be developed by either a top-down or bottom-up approach and from either statistical analysis, surveys, theoretical calculations, pilot projects or policy objectives [4].

In the literature, these benchmarks were typically calculated via quartile ranges. Ganassali et al. [19], Simonen et al. [10], and Paratscha et al. [29] defined the limit, reference, and target boundary ranges according to the higher, inter, and lower interquartile ranges. Meanwhile, Rucińska et al. [30] used the standard deviation, and Hollberg et al. [20] set more competitive target values corresponding to the 0.05 percentile of the available solutions.

For the purpose of this TG, these values were selected based on the literature mentioned above using percentiles but taking less strict values as a first approach to obtain benchmarks for asphalt mixtures. Therefore, the limit value was set as the 0.8 percentile of the values obtained, and the short-term target value was chosen as the 0.10 percentile of the values obtained. The long-term target value was not calculated for being out of the scope of this investigation.

Once the three benchmarks (i.e., reference, limit, and target) for each impact category indicator were calculated using the conventional asphalt mixtures case studies with 50 km as transport distances, they could be compared with the available case studies of asphalt mixtures containing waste and secondary materials (WMR case

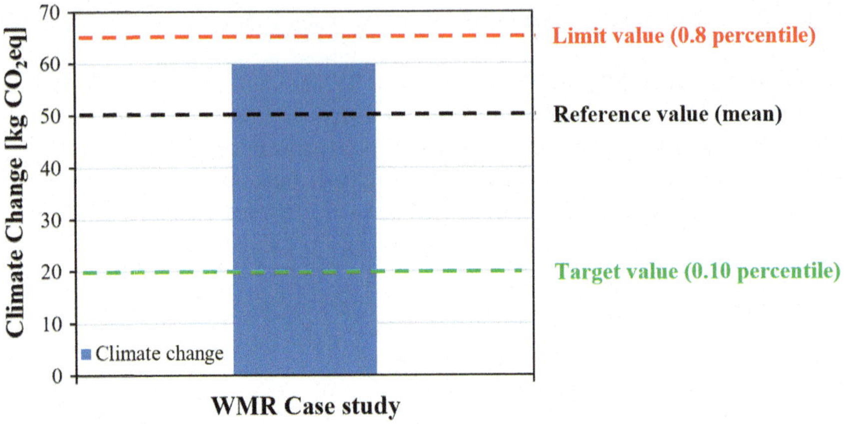

Fig. 17 Hypothetical benchmark analysis

studies). Figure 17 shows how this average would be made. In this example, the impact indicator Climate Change of the case study has a higher value than the reference, meaning that it has more impact than the average but is still lower than the limit we established. The impact is quite far from the target of the 0.10 percentile.

2.2 *Environmental Impact of Asphalt Mixtures with WMR*

In order to assess the environmental impact of asphalt mixtures containing waste and secondary materials (WMR mixtures), the following steps were followed:

1. Collecting WMR case studies from partners of TG3 and TG4
2. Performing LCA of WMR case studies
3. Comparison of WMR case studies with the generated benchmarks.

These steps are described in the following sections.

2.2.1 Data Collection

In order to collect WMR case studies, the "RILEM TC WMR TG4 LCA Data Collection Tool" was used to collect WMR case studies in the same way as for conventional asphalt mixtures. Of note, only two case studies of asphalt mixtures containing crumb rubber were received.

2.2.2 LCA of WMR Case Studies

The LCA of WMR case studies was performed following the four steps defined in ISO standards 14,040 [23] and 14,044 [25] and following the main four steps, as follows:

Goal and Scope Definition

The aim of this LCA is to calculate the environmental impacts of different WMR mixtures during their product stage (A1-A3) to be compared with the benchmarks.

The intended applications of the study are:

- The understanding of environmental burdens due to the production of WMR asphalt mixtures;
- The support of a more sustainable decision-making process in the asphalt industry.

All considered mixes were produced in plants from different parts of the world and all the related information was provided by the specific TG partner. Three life cycle stages were analysed, namely: Raw Material Supply (A1), Transport to Plant (A2), and Manufacturing (A3) (Fig. 12). Hence, the analysis performed is a so-called "cradle-to-gate": from the extraction of raw materials ("cradle") to the product ready to be used and distributed ("gate" of the asphalt plant). These life cycle stages correspond to the life cycle of asphalt mixtures as road pavement components before a service life or durability is assigned to them. In this regard, the analysis performed is focused on the materials, their obtention and their design.

As required by the standards, a declared unit was chosen: 1t of asphalt mixture to be produced. Concerning the cut-off rules, none was applied in order to consider the whole material impact. The selected impact categories are related to Energy Consumption and Climate Change, and they were calculated through the EF 2.0 methodology, developed by the European Commission in 2018. Concerning the data, they have to comply with some requirements [4]. This is detailed in the Life Cycle Inventory (LCI) section.

Life Cycle Inventory

This step in the LCA is strictly connected to the data collection, performed using the "RILEM TC WMR TG4 LCA Data Collection Tool". Most data were directly provided by the TG partners (primary data). When not available, GaBi ts database was used.

Two case studies, including Crumb Rubber (CR) were analysed. CR was modelled in GaBi ts using the information provided by an Italian LCA. Table 3 show the mix design, and Table 4 presents the transport and energy used in each case study. In the system boundaries of these case studies, the production of CR was considered

Table 3 Mix design of each case study

Case study	Fine aggregates		Coarse aggregates		Filler		Binder	
	Type	Percentage in the mix	Type	Percentage in the mix	Type	Percentage in the mix	Type	% in the mix
EMPA_SDA with CR	–	–	–	93.8	–	–	PMB CR	6.20
UGR_AC with CR	Crushed rock-limestone	51.5	Crushed rock-limestone	38.0	CEM I	5.5	PMB CR	4.2

including the amount of water, electricity and diesel needed for the transport of waste tyres within the plant, which was given by a previous LCA.

Life Cycle Impact Assessment

Same method as Sect. 2.1.4.3.

Interpretation of Results

Same method as Sect. 2.1.4.4.

3 Results

The results section is divided into two sections according to the aims of the TG, hence presenting firstly the results of the LCA benchmarking of asphalt mixtures, and secondly, the results of the LCA of asphalt mixtures containing WMR being compared with the benchmarks.

Table 4 Transport and energy used of each case study

Case study	Transport of fine aggregates		Transport of coarse aggregates		Transport of filler		Transport of binder		Energy used	
	Type	Distance (km)	Type	Distance (km)	Type	Distance (km)	Type	Distance (km)	Type	Amount
EMPA_SDA with CR	Euro V	118	Euro V	118	Euro V	118	Euro V	311	Electricity Natural gas	8.64 kWh/t 6.17 m^3/t
UGR_AC with CR	Euro V	30	Euro V	30	Euro V	30	Euro V	30	Electricity Diesel	23.98 kWh/t 6.98 l/t

3.1 LCA Benchmarking for Asphalt Mixtures

3.1.1 Results of the LCA of Conventional Asphalt Mixtures Case Studies

According to the steps defined for the LCA benchmarking of asphalt mixtures, the LCA of different case studies for reference asphalt mixtures was carried out for AC and SMA using one tonne of asphalt mixture as the declared unit. The results presented here are therefore referred to one tonne of asphalt mixture produced.

Figures 18 and 19 show the results of the climate change and energy use indicators, respectively, for the 14 case studies with real distances. The impact indicators are divided into the three life cycle stages (A1: raw material supply, A2: transport to plant, A3: manufacturing). In addition, the different case studies are displayed from minimum binder content (left) to maximum binder content (right), showing that energy use increases with increasing bitumen content.

Regarding Climate Change (Fig. 8), the impact of raw materials supply (A1) tends to increase with the bitumen content. The A1 climate change emissions go from 39.97 kg CO_2 eq for TNO_Open Stone with 2.92% of bitumen to 55.99 kg CO_2 eq with 7.30% of bitumen for TNO_SMA8-11. TNO_Open Stone and TNO_AC SURF disrupt the increasing trend of global warming potential due to the significantly higher transport distances and including transport by ship. Moreover, the high emissions of UNIPA_AC32TS are due to the amount of energy (electricity and oil) for production. It can also be stated that UAW_AC20 has a lower impact due to the use of limestone, which impacts less than other virgin materials during the extraction and due to less quantity of energy used for the mixing. UVS_AC11 has a lower contribution in terms of climate change because of the shorter transport distances compared to the other

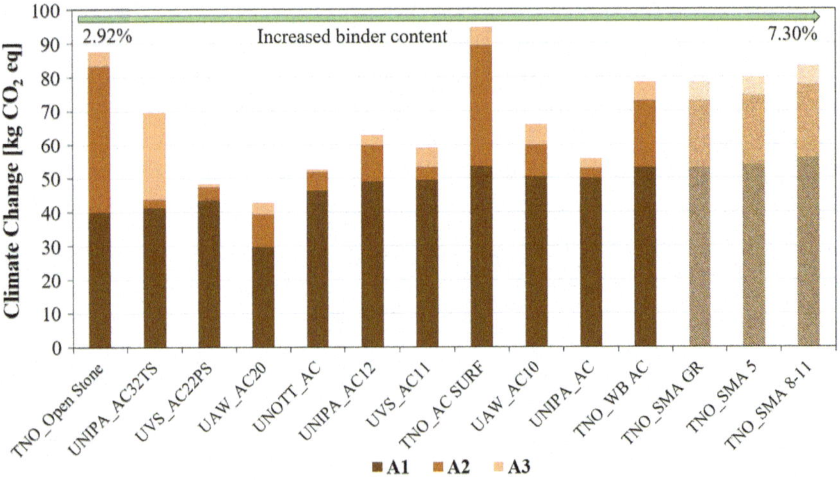

Fig. 18 Climate change of conventional asphalt mixtures with contribution of each life cycle stage

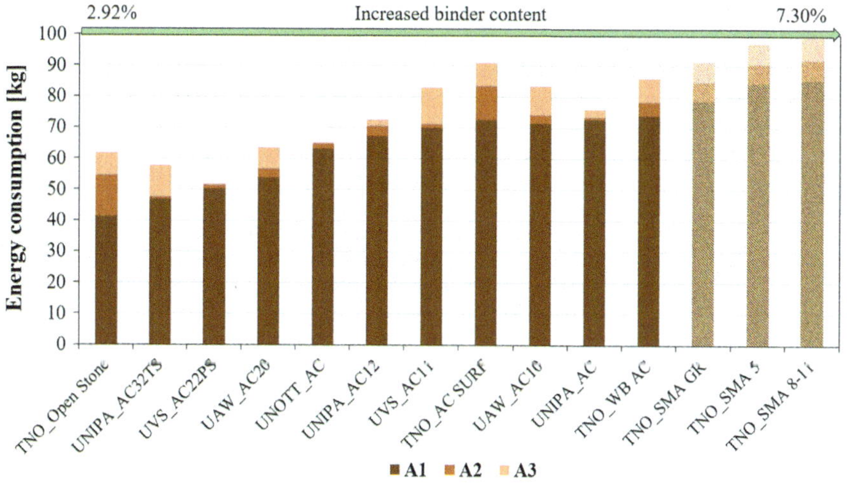

Fig. 19 Energy Consumption of conventional asphalt mixtures with contribution of each life cycle stage

mixtures. A similar observation can be made for UNIPA_AC, with a relatively lower impact due to the reduced energy used in mixing. The range for climate change emissions goes from 42.90 to 94.80 kg of CO_2 eq for AC, while it varies from 78.80 to 83.35 kg CO_2 eq for SMA.

Concerning the Energy Use, it is clear that the raw materials supply influences this indicator, and in particular, it increases with the increase in bitumen content, while it is less affected by transport and mixing (Fig. 19). The range for energy use goes from 61.83 to 85.96 kg of energy for AC and from 91.50 to 99.25 kg for SMA.

Regarding mixture type, it can be observed that SMA mixtures present higher environmental impact than AC mixtures, probably due to the higher binder content and type. This is true for both Energy Used and Climate Change, although only for those case studies amongst the collected ones not including materials supplying through boat shipping (case studies from TNO).

Figures 20 and 21 show the results of the hotspot analysis performed to identify the most significant life cycle stage in each case study. It can be observed that:

- For most case studies, A1 is a hotspot or the most significant stage, regardless of the type of mixture, contributing from 56 to 90% for both Climate Change and Energy Use.
- A2 contributes from 4 to 49%. It is evident that the mixtures where A2 has a higher weight are those whose aggregates are transported both by truck and by ship, which is the case of TNO (Netherlands) asphalt mixtures. In particular, Porous Asphalt records the highest contribution (49%) due to the transport of coarse aggregates from Scotland.

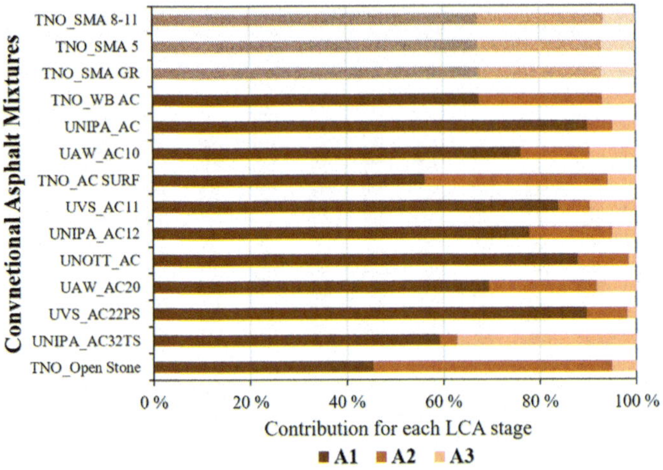

Fig. 20 Hotspot analysis—climate change for conventional mixtures

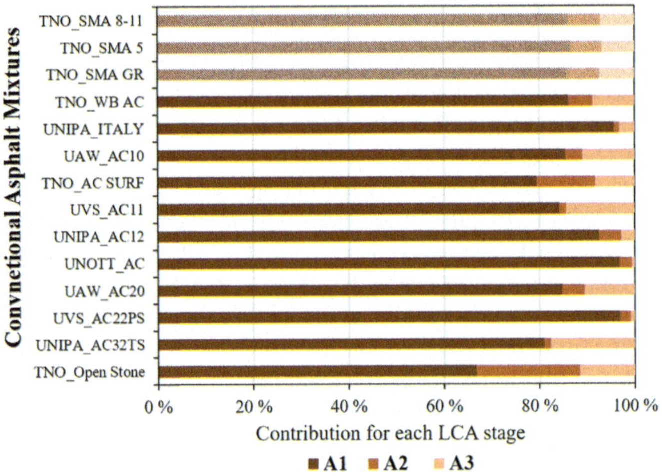

Fig. 21 Hotspot analysis—energy consumption for conventional asphalt mixtures (AC and SMA)

- In general, A3 has the lowest impact. The highest contribution is recorded for those mixtures which require electricity+natural gas (i.e., UVS_AC11) or electricity+heating oil (i.e., UNIPA_AC32TS).

Tables 5 and 6 show the sensitivity analysis results carried out for transport distances. All transport distances in Table 5 were set as 50 km for road transport by truck and 600 km for inland shipping, while the results in Table 6 assumed 150 km for road transport by truck and 1000 km for inland shipping. The results shown are the difference in percentage obtained when comparing the assumed distances with the actual distances. The following observations were made:

Table 5 Sensitivity Analysis of conventional mixtures with distances assumed 50 km for trucks and 600 km for inland shipping. CC = climate change

Istitution	Country	Year	Energy (kg) (%)	CC (CO_2 eq) (%)
TNO_Open Stone	Netherlands	2020	−14	−32
UNIPA_AC32TS	Germany	2020	1	2
UVS_AC22PS	Lithuania	2021	0	−1
UAW_AC20	Belgium	2019	−3	−13
UNOTT_AC	UK	2015	−1	−4
UNIPA_AC12	Republic of San Marino	2020	−3	−11
UVS_AC11	Lithuania	2021	−1	−3
TNO_AC SURF	Netherlands	2020	−7	−20
UAW_AC10	Belgium	2019	−2	−8
UNIPA_AC	Netherlands	2020	0	1
TNO_WB AC	Italy	2015	0	1
TNO_SMA GR	Nctherlands	2020	−2	−7
TNO_SMA 5	Netherlands	2020	−2	−8
TNO_SMA 8-11	Netherlands	2020	−2	−8

Table 6 Sensitivity analysis of conventional mixtures with distances assumed 150 km for trucks and 1000 km for in-land shipping. CC = climate change

Istitution	Country	Year	Energy (kg) (%)	CC (CO_2 eq) (%)
TNO_Open Stone	Netherlands	2020	−6	−14
UNIPA_AC32TS	Germany	2020	5	13
UVS_AC22PS	Lithuania	2021	26	28
UAW_AC20	Belgium	2019	1	5
UNOTT_AC	UK	2015	3	10
UNIPA_AC12	Republic of San Marino	2020	0	1
UVS_AC11	Lithuania	2021	2	9
TNO_AC SURF	Netherlands	2020	−1	−3
UAW_AC10	Belgium	2019	1	3
UNIPA_AC	Netherlands	2020	6	22
TNO_WB AC	Italy	2015	3	15
TNO_SMA GR	Netherlands	2020	3	13
TNO_SMA 5	Netherlands	2020	3	10
TNO_SMA 8-11	Netherlands	2020	3	10

- When the transport distances are set as 50 km for trucks and 600 km for ships (Table 5), both impact indicators are reduced for most case studies. When the transport distances are set as 150 km for trucks and 1000 km for ships (Table 6), both impact indicators are increased for most case studies.
- Climate change seems to be more influenced by transport distances than Energy Use. When transport distance reduces, Climate Change lessens more than the energy use indicator (hence, of diesel), as observed in Table 5. This is due to the higher impact that A2 has for Climate Change than for Energy Use, as observed in the hotspot analysis in which A2 represents a higher percentage of the total for all case studies.
- This sensitivity analysis demonstrates the impact that transport distances may have in the life cycle of an asphalt mixture:
  - When the distances are reduced by approximately 50%, a reduction of 32% in climate change and 14% in energy use in the case of TNO_Open Stone (Table 5) are obtained.
  - When distances are increased by 57% (Table 6), an increase of 28% in Climate Change and 26% in energy use (UVS_AC22PS) is obtained.

3.1.2 Benchmark Generation

The final step of the benchmarking methodology consists of creating the benchmarks from the conventional asphalt mixtures case studies.

Since the sensitivity analysis showed that transport distances of 50 km for trucks and 600 km for inland shipping would not significantly alter the impact of A2 stage, the case studies with these distances were used to generate the benchmarks. In this way, the benchmarks are generated from a homogeneous database, using constant transport distances and avoiding having deviations from particular case studies, such as those of TNO.

Table 7 shows the benchmarks obtained following the methodology described in Sect. 2.1.5 for each impact indicator and each type of asphalt mixture. The benchmarks for SMA mixtures are higher than those for AC mixtures due to the higher environmental impact of SMA mixtures. These benchmarks should be used as following:

- The reference values are the average of all case studies for each type of mixture and indicator. Thus, these values should be compared with WMR case studies to determine whether the latter has a higher or lower environmental impact than reference asphalt mixtures.
- The limit values are those for which 80% of case studies for conventional asphalt mixtures have an inferior value, and therefore is set as the maximum de-sired impact not to be surpassed.
- The target values are exceeded by the majority of case studies with conventional asphalt mixtures (90%), and therefore, they are set as the target to be obtained.

Table 7 Generated benchmarks for climate change and energy use indicators for asphalt mixtures (cradle-to-gate) All values are referred to one ton of asphalt mixture produced

	Reference		Limit		Target	
	Climate change ($kgCO_2eq$)	Energy use (kg)	Climate change ($kgCO_2eq$)	Energy use (kg)	Climate change ($kgCO_2eq$)	Energy use (kg)
Asphalt concrete	57.50	68.04	60.40	82.96	48.26	51.84
Stone mastic asphalt	74.58	94.12	76.72	97.18	73.38	89.87

It is worth mentioning that the benchmarks have been generated from 11 case studies in the case of asphalt concrete and three case studies for stone mastic asphalt. In this regard, the representatively of these values would increase when the number of case studies increased.

3.2 *Environmental Impact Assessment of Asphalt Mixtures with WMR*

The results of the climate change and energy use indicators of the WMR case studies are shown in Figs. 22 and 23 divided into life cycle stages and together with the benchmarks obtained for reference asphalt mixtures. The first noticeable observation is that the two CR mixtures exhibit very different Climate Change values due to the A1 stage being higher for the UGR mix, while they show more similar values for Energy Use. For both case studies, A1 is the most impactful stage, particularly for UGR mixture. The reason lies in the mix design; in fact, the use of Cement (CEM I) as filler contributes more than half of the total amount of CO_2 eq.

Regarding Climate Change, UGR_AC with CR overpasses the benchmarks for AC, but not the ones for SMA, revealing that the type of asphalt mixture is key in assessing its environmental impact. On the other hand, the mixture from EMPA (SDA, a semi dense asphalt mixture with high binder content (6%) and high air voids (12–18%)) is more environmentally friendly than conventional asphalt mixtures and shows values that are below the target of AC. The difference in the values of Climate Change between UGR_AC and EMPA_SDA is due to the different materials (A1) used in the mixtures, and mainly related to the use of cement as filler in UGR_AC. Concerning Energy Use, both case studies remain below the reference benchmark for AC. This means that their environmental impact is lower than that of the selected reference mixtures.

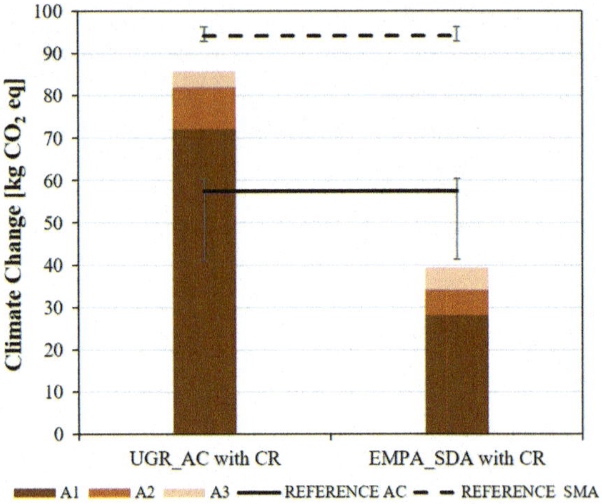

Fig. 22 Climate change indicator for WMR case studies compared to the benchmarks with reference target and limit

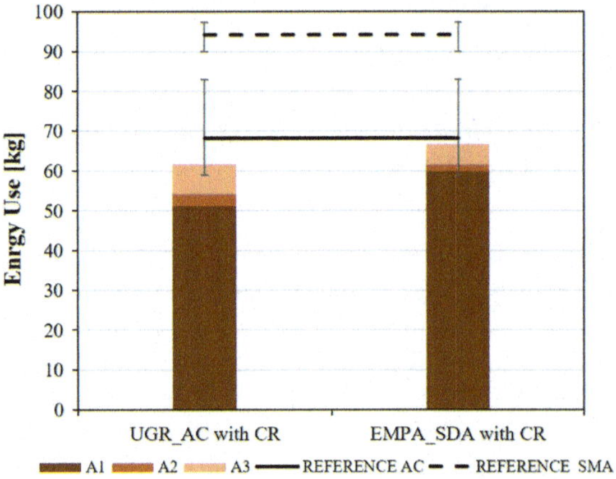

Fig. 23 Energy use indicator for WMR case studies compared to the benchmarks with reference target and limit

4 Summary, Conclusions and Future Recommendations

In order to foster the implementation of more sustainable practices in the paving industry, this study provides environmental impact benchmark values (Fig. 24), in terms of energy use and climate change, of conventional asphalt mixtures and asphalt mixture containing WMR.

To achieve this goal, a comprehensive collection of primary data and case studies from the TG partners was undertaken to populate the Life Cycle Assessment (LCA) database. The first step focused on defining benchmarking values of LCA results for reference asphalt mixtures. With regards to the LCA of conventional asphalt mixtures, and subsequent benchmarking, it was decided to divide the case studies into two subgroups: Asphalt Concrete (AC) and Stone Matrix Asphalt (SMA). In addition, during this step, the case studies collected were also investigated by means

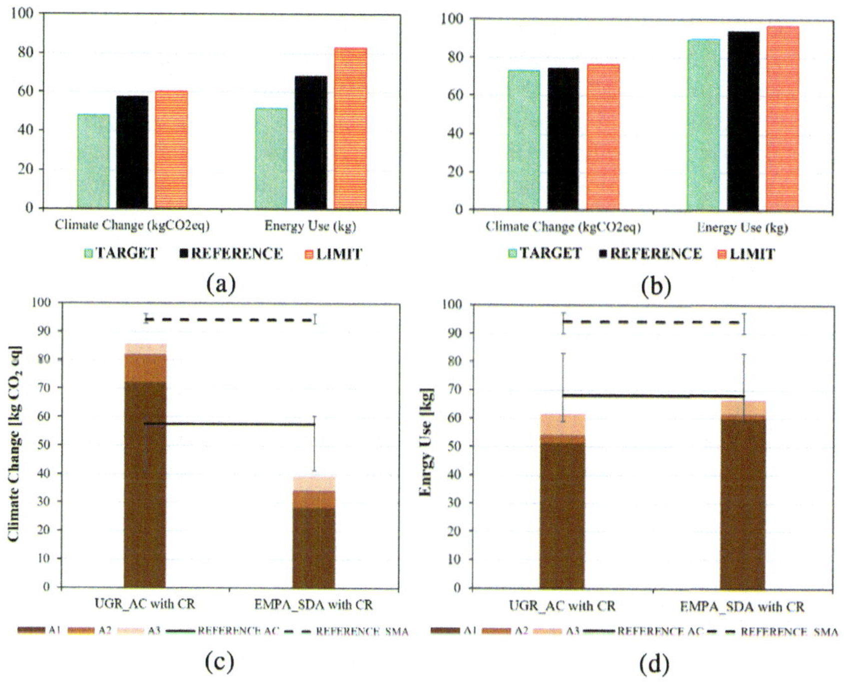

Fig. 24 **a** Generated benchmarks for climate change and energy use of asphalt concrete mixtures (cradle-to-gate); **b** generated benchmarks for climate change and energy use of stone mastic asphalt mixtures (cradle-to-gate); **c** climate change indicator for WMR case studies compared to the benchmarks with reference target and limit; **d** energy use indicator for WMR case studies compared to the benchmarks with reference target and limit

of hotspot and sensitivity analyses. Secondly, to assess whether including waste and marginal materials is a sustainable practice, the LCA results of two additional case studies of asphalt mixtures with WMR were compared with the benchmarks obtained.

As a result, the following is a summary of results:

- Conventional asphalt mixtures are influenced by the amount of bitumen. Hence, if bitumen increases, climate change and energy use also increase. Nevertheless, this trend is not very clear if transport distances are particularly long, in which case they can influence the total result, making it higher, independently of the quantity of bitumen.
- As general conclusion, Climate Change seems to be more influenced by transport distances, while energy consumption is more dependent on mix design.
- Due to the higher bitumen content and type, SMA mixes consume more energy and have a higher impact on climate change than ACs.
- The hotspot analysis performed showed that for a cradle to gate analysis:
  - A1 (raw material supply) is always the most impactful stage, regardless of the mixture investigated.

  - Concerning A2, transport distances influence Climate Change more than Energy Consumption. A2 contributes from 4 to 49% in terms of climate change, while from 1 to 22% in terms of energy consumption.
  - Concerning A3, it is never a hotspot or a relevant stage in the cradle to gate analysis. Its contribution, in terms of energy consumption and climate change, increases if the mixing plant requires both electricity and natural gas/heating oil.

- The sensitivity analysis on transport distances showed that:

  - Climate Change is more sensitive to transport distances than Energy Use. This is due to the higher impact that A2 has for Climate Change than for Energy Use, as observed in the hotspot analysis in which A2 represents a higher percentage of the total impact for all case studies.
  - If transport distances are a relevant life stage in the life cycle of an asphalt mixture, a reduction of 50% of such distances can lead to a reduction of 32% in Climate Change and 14% in Energy Use.

- Three benchmarks have been generated from the conventional asphalt mixtures case studies collected:

  - The reference values are the average of all case studies for each type of mixture and indicator, so it is the value to compare with WMR case studies to understand whether their impact is greater or lower than that of reference asphalt mixtures.
  - The limit values are those for which 80% of case studies with conventional asphalt mixtures have an inferior value, and therefore is set as the maximum desired impact not to be surpassed
  - The target values are those for which the 90% of case studies for conventional asphalt mixtures have a superior value, and therefore is set as the target to be obtained.
  - The benchmarks for SMA mixtures are higher than those for AC mixtures due to the higher environmental impact of SMA mixtures.
  - The methodology employed to develop benchmarks may serve as an initial point to compare with WMR asphalt mixtures. However, these values should come from a larger number of case studies.

- The analysis of the WMR case studies and comparing them to the benchmark show that these values are very useful to make direct comparisons and quickly obtain conclusions about the environmental impact of WMR technologies. In this regard:

  - In terms of Climate Change, the type of conventional mixture makes the difference. In fact, in any case rubberized asphalt mixtures are more environmentally friendly than conventional SMA. Instead, when compared with the AC benchmark values, there is not a clear trend: this is probably due to the presence of cement within one of the mixtures (UGR_AC with CR).
  - On the other hand, WMR case studies show equivalent or less energy use than conventional asphalt mixtures. Specifically, SMA always consumes a

higher amount of energy, while energy consumption of AC is comparable with benchmarks of conventional mixtures.

Keeping in mind that the analysis is limited to A1–A3 (cradle-to-gate), the following conclusions are made:

- Raw materials supply is the most relevant life cycle stage for Climate Change and Energy Use, unless transport distances are particularly long (only observed for TNO case studies including transport by ship). In this regard, if the environmental impact of asphalt mixtures is to be decreased, practitioners should focus on reducing the impact of the extraction of raw materials or focus on increasing the use of recycled materials in asphalt mixtures, however, this was not part of this study.
- SMA mixtures show higher environmental impact than AC mixtures in terms of materials. Therefore, although not part of this study, it can be speculated that designers should try to use more recycled materials in SMAs and/or possibly compensate with greater durability, considered in the life cycle of asphalt pavements.
- Selected WMR mixtures seem to be a more environmentally friendly option when compared to benchmark values of SMA. This is not always as clear when the selected WMR mixes are compared to conventional ACs.

On the basis of these finding the authors believe necessary providing the following recommendations for future studies:

- The use of benchmarks is a very practical tool for decision making. However, the generation of these benchmarks should come from a large number of case studies of either conventional asphalt mixtures or WMR asphalt mixtures. Future studies should then consider the reported findings limited to the selected case studies and should certainly try using the same methodology but enlarging the statistical population of case studies and possibly consider using a probabilistic approach.
- The asphalt industry should focus on collecting data for LCA using a standardized procedure. This TG recommends using the "RILEM TC WMR TG4 LCA Data Collection Tool" which was specifically created for this purpose and to ease this task. The tool developed in this work is readily available for further use by requesting the following authors: davide.lopresti@unipa.it; ajbc@ugr.es. This study is limited to A1–A3 (cradle-to-gate) which are believed to be the correct system boundaries to assess the environmental impact of asphalt mixtures towards the definition of Environmental Product Declaration. However, future studies should consider also providing recommendations for the recyclability of the products (module D).

References

1. Häkkinen T, Antuña C, Mäkeläinen T, Lützkendorf T et al (2012) Sustainability and performance assessment and benchmarking of buildings. Final report
2. Stripple H. Life cycle assessment of road: a pilot study for inventory analysis. http://www.ivl.se. Accessed 08 Feb 2019
3. Harvey JT, Meijer J, Ozer H, Al-Qadi IL et al. Pavement life cycle assessment framework. https://www.fhwa.dot.gov/pavement/sustainability/hif16014.pdf. Accessed 10 Dec 2019
4. ISO 21678:2020 (2020) Sustainability in buildings and civil engineering works—indicators and benchmarks—principles, requirements and guidelines. International Organization for Standardization
5. European Commission, Welcome to Intelligent Energy Europe (IEE)—European Commission. https://ec.europa.eu/energy/intelligent/projects/. Accessed 04 Sep 2020
6. Bowick M, O'connor J, Meil J. Whole-building LCA benchmarks: a methodology white paper. https://www.athenasmi.org. Accessed 10 Dec 2021
7. ALCAS, The Australian life cycle inventory database initiative. http://www.auslci.com.au/. Accessed 13 Oct 2020
8. HQE Association, Construisons Ensemble 'HQE Performance'—Bilan deux and apres le lancement de la charte d'engagement. http://www.assohqe.org. Accessed 13 Oct 2020
9. OneClickLCA, Carbon Heroes Benchmark Program. https://www.oneclicklca.com/construction/carbonheroes/. Accessed 03 Sep 2020
10. Simonen K, Rodriguez B, Barrera B, Huang M et al. Embodied carbon benchmark study: LCA for low carbon construction—part one. https://digital.lib.washington.edu/researchworks/handle/1773/38017. Accessed 03 Sep 2020
11. Röck M, Saade MRM, Balouktsi M, Rasmussen FN et al (2020) Embodied GHG emissions of buildings—the hidden challenge for effective climate change mitigation. Appl Energy 258:114107. https://doi.org/10.1016/j.apenergy.2019.114107
12. German Federal Ministry of the Interior, B. and C., Ökobaudat Home Page: Assessment System for Sustainable Building (BNB). https://www.oekobaudat.de/en/home/assessment-system-for-sustainable-building-bnb.html. Accessed 13 Oct 2020
13. BREEAM, Building LCA benchmarks, BREEAM UK New Construction 2018. https://www.breeam.com/news/building-lca-benchmarks-breeam-uk-new-construction-2018/. Accessed 13 Oct 2020
14. Schlegl F, Gantner J, Traunspurger R, Albrecht S et al (2019) LCA of buildings in Germany: proposal for a future benchmark based on existing databases. Energy Build 194:342–350. https://doi.org/10.1016/j.enbuild.2019.04.038
15. Saboori A, Harvey JT, Butt AA, Jones D (2017) Life cycle assessment and benchmarking of end of life treatments of flexible pavements in California. In: Pavement LCA 2017 Symposium. https://lcasymposium.ict.illinois.edu/files/2018/01/72_Pvmt-LCA-2017-EOL-Benchmarking-clean-1.pdf. Accessed 13 June 2020
16. Butt AA, Harvey JT, Saboori A, Ostovar M et al (2019) Airfield life cycle assessment: Benchmark study of a project at JFK International Airport. In: Airfield and highway pavements 2019: innovation and sustainability in highway and airfield pavement technology—selected papers from the international airfield and highway pavements conference 2019. American Society of Civil Engineers (ASCE), pp 456–464. https://doi.org/10.1061/9780784482476.045
17. NAPA, General Program Instructions for Environmental Product Declarations (EPD) Program. http://www.asphaltpavement.org/PDFs/EPD_Program/NAPA_GeneralProgramInstructions_FINAL.pdf. Accessed 13 Mar 2009
18. Muench ST, Migliaccio G, Kaminsky J, Ashtiani MZ et al (2019) NCHRP research report 916: Sustainable highway construction guidebook. Washington DC, USA. https://doi.org/10.17226/25698
19. Ganassali S, Lavagna M, Campioli A., Saporetti S (2018) Green public procurement and construction sector: EPD and LCA based benchmarks of the whole-building. In: Designing sustainable technologies, products and policies. Springer, Cham, pp 503–513

20. Hollberg A, Vogel P, Habert G (2019) LCA benchmarks for decision-makers adapted to the early design stages of new buildings. In: Life-cycle analysis and assessment in civil engineering: towards an integrated vision—Proceedings of the 6th international symposium on life-cycle civil engineering, IALCCE 2018. CRC Press/Balkema, pp 775–781
21. De Wolf C, Davies D (2019) Benchmarking embodied carbon in structural materials. Life-cycle analysis and assessment in civil engineering: towards an integrated vision—Proceedings of the 6th international symposium on life-cycle civil engineering, IALCCE 2018, pp 2965–2972
22. Pelkmans L, Devriendt N, Junginger M, Hoefnagels R et al (2012) Benchmarking biomass sustainability criteria for energy purposes. Final report
23. ISO 14040:2006 (2006) Environmental management—life cycle assessment—principles and framework. International Organization for Standardization
24. Moins B, Hernando D, Buyle M, France C, Audenaert A (2022) On the road again! An economic and environmental break-even and hotspot analysis of reclaimed asphalt pavement and rejuvenators. Resour Conserv Recycl 177:106014. https://doi.org/10.1016/j.resconrec.2021.106014
25. ISO 14044:2006 (2006) Environmental management—life cycle assessment—requirements and guidelines. International Organization for Standardization
26. PavementLCM (2021) Lo Presti D et al Deliverable D5.1a—pavement LCM guidelines. https://www.pavementlcm.eu/pavementlcm-guidelines/. Accessed 14 Mar 2022
27. Hernando D, Moins B, Van den bergh W, Audenaert A (2022) Identification of the main environmental impact categories over the life cycle of hot mix asphalt: an application to green public procurement. Transp Res Rec 03611981221083616. https://doi.org/10.1177/036119812 2108361
28. European Commission—Joint Research Centre, Guide for interpreting life cycle assessment result. https://publications.jrc.ec.europa.eu/repository/bitstream/JRC104415/lb-na-28266-en-n.pdf. Accessed 14 May 2022
29. Paratscha R, Von Der Thannen M, Smutny R, Lampalzer T et al (2019) Development of LCA benchmarks for Austrian torrent control structures. Int J Life Cycle Assess 24:2035–2053. https://doi.org/10.1007/s11367-019-01618-6
30. Rucińska J, Komerska A, Kwiatkowski J (2020) Preliminary study on the GWP benchmark of office buildings in Poland using the LCA approach. Energies 13(13):3298. https://doi.org/10.3390/en13133298

Chapter 6
Conclusions

Augusto Cannone Falchetto, Di Wang, Lily D. Poulikakos, and Emiliano Pasquini

Abstract The final chapter of this STAR document provides a closing synthesis of the activities of the TC 279-WMR on the Valorisation of Waste and Secondary Materials for Roads. A brief overview of the technical committee is followed by a summary of the most significant outcomes from each of the four active Task Groups. Finally, the research needs identified at the end of the TC activities are presented, and a perspective for continuing and expanding the scientific research on alternative materials for roads is proposed.

Keywords Polyethylene · Crumb rubber · Alternative aggregates · Recyclability · LCA · Environmental performance

1 Overview

The last chapter of this State-of-the-Art (STAR) Report summarises the results of the journey toward a new direction in the research area in the technical committees of RILEM as accomplished by TC 279-WMR on Valorisation of Waste and Secondary Materials for Roads. In addition, the final section of the present chapter is devoted to presenting a synthesis of the current research needs from the perspective of advancing the effort of TC 279-WMR, which laid the foundation for new research on alternative paving materials.

A. Cannone Falchetto (✉) · D. Wang
Department of Civil Engineering, Aalto University, Rakentajanaukio, 4-02150 Espoo, Finland
e-mail: augusto.cannonefalchetto@aalto.fi

L. D. Poulikakos
EMPA, Swiss Federal Laboratory for Material Science and Technology, Überlandstrasse 129, 8600 Dübendorf, Switzerland

E. Pasquini
Department of Civil, Environmental and Architectural Engineering (ICEA), University of Padova, Via Marzolo 9, 35131 Padova, Italy

© The Author(s), under exclusive license to Springer Nature Switzerland AG 2023
A. Cannone Falchetto et al. (eds.), *Valorisation of Waste and Secondary Materials for Roads*, RILEM State-of-the-Art Reports 38,
https://doi.org/10.1007/978-3-031-33173-2_6

The large amount of unused, unrecycled, and often disposed materials derived from diverse activities and industries and their valuable potential when incorporated into pavement materials was the basis for conceiving and developing this novel RILEM technical committee. The organization into task groups covering materials that could be used as additives, modifiers, and aggregates demonstrated the attempt of TC 279-WMR to comprehensively address several factors associated with the use of waste and secondary materials for pavement applications. Suitability, characterisation, rheological behaviour, mechanical response, and environmental sustainability are essential for evaluating the possibility of using these alternative materials.

In view of the international cooperation characterising the RILEM activities and to facilitate extensive participation from a diverse group of members in the experimental phase of the research, four Task Groups (TGs) were active, as described in detail in Chap. 1 of this STAR document:

- TG1: Asphalt Binder and Asphalt Mixture Additives working on waste plastic and specifically on Polyethylene
- TG2: Asphalt Binder Additives addressing the use of crumb rubber obtained from vehicle tires
- TG3: Aggregate and Filler Substitutes investigating materials such as construction and demolition waste, recycled concrete aggregates, and steel slag; and cooperating with TG1 on the incorporation of waste polyethylene in asphalt mixtures
- TG4: Life Cycle Assessment (LCA) estimating the environmental impact of the paving materials containing waste and secondary materials based on LCA techniques.

2 Synthesis on the Outcome of the Technical Committee

In this section, the results of each of the four active Task Groups (TGs) are synthetically presented as a summary of the outcome of the entire Technical Committee. It is worth remarking that the work performed in the TC 279-WMR was the first attempt in the RILEM community to address the use of waste and secondary materials for paving purposes. Hence, the results obtained during the intense research activity must be considered exploratory while establishing the basis for a new research direction.

2.1 Task Group 1

Task Group 1 evaluated the incorporation of waste Polyethylene (PE) into binder blends and mixtures based on extensive interlaboratory experimentation with a shared testing protocol and a comprehensive set of tests. Blends of binder and PE pellets or shreds were characterized based on conventional and rheological testing and chemical analysis. Asphalt mixtures containing PE additives were also prepared, and their mechanical properties were evaluated in collaboration with TG3. The following key points can be highlighted from the TG1 activities:

- The PE-modified binder blends showed higher resistance to permanent deformation and better fatigue performance than the original binder as long as the strain level was not triggering any potential brittle phenomena.
- The PE-modified mixtures exhibited unaltered workability and improved Indirect Tensile Strength (ITS) response, given the enhanced cohesion associated with the plastic-modified mastic phase. Substantial benefits were also observed regarding stiffness and creep response (rate and modulus), further confirmed by the wheel tracking test. Comparable or better moisture resistance was also found for PE-modified materials, supporting the promising potential for using PE waste for paving purposes.

2.2 *Task Group 2*

Task Group 2 addressed the use of Crumb Rubber (CR) and its potential as a performance-enhancing binder additive. The CR from end-of-life tires obtained from four manufacturing processes was blended with binders with three distinct grades. Conventional, rheological, chemical properties and microscope imaging was used to investigate the different combinations of binder and CR. Given the results obtained from the interlaboratory work, it was observed that:

- The incorporation of CR affects the overall stiffness spectrum of the original binder with beneficial higher stiffness at high temperatures (34–82 °C) and a reduction of the same property at low temperatures (0–40 °C). This behaviour is mainly associated with a physical interaction between CR and binder, as confirmed by the FTIR analysis, with the grade of the binder playing a fundamental role in determining the response of the CR-modified binder.

2.3 *Task Group 3*

Task Group 3 explored different waste materials as alternative aggregates for asphalt mixtures. These materials included Construction and Demolition Waste (CDW), Recycled Concrete Aggregates (RCA), and Steel Slag (SS). A peculiar aspect of this work was selecting local secondary aggregate material and designing the mixtures. The following considerations are proposed based on the experimentation conducted by the TG3 members:

- SS exhibited better characteristics among the three alternative aggregates, with CDW and RCA presenting poorer properties than the natural aggregates, such as limestone and sandstone.

- While more binder was needed for designing mixtures with waste aggregates, they showed a superior mechanical response, including higher resistance to permanent deformation and a positive impact on the tensile stiffness modulus.
- In view of the equivland or improved performance observed, turning the recycling of SS, CDW, and RCA into common practice would have multiple benefits. The list includes reducing the amount of waste, saving virgin aggregates, lowering the production costs, promoting conscious and virtuous economic and environmental practices, and limiting the impact of the paving construction industry on the environment and society.

2.4 *Task Group 4*

Task Group 4 investigated alternative paving materials to provide benchmark values of the environmental effect of conventional asphalt mixes, and asphalt mixtures incorporating waste and secondary materials to encourage the adoption of more environmentally friendly methods in the paving sector. The focus of the performed Life Cycle Analysis (LCA, from cradle-to-gate) was on two key parameters, Energy Use and Climate Change, for the two types of mixtures analysed: Asphalt Concrete (AC) and Stone Matrix Asphalt (SMA). Hotspot and sensitivity analyses were also performed to evaluate the sustainability of incorporating byproducts into asphalt mixtures. Based on the proposed research approach, the following key results were obtained:

- Mix design and transport distance mainly affect energy use and climate change, respectively, with SMA having the higher impact due to its higher binder content. Based on the hotspot analysis, the raw material supply phase is the most critical overall, followed by transport distances, while a minor contribution to energy consumption and climate change is provided by the manufacturing phase, which never appears to be a hotspot. Additionally, the sensitivity study revealed that energy use is less sensitive to transportation distances than climate change.
- Based on the three benchmark levels defined (reference, limit, and target values), AC exhibits a higher environmental impact than SMA mixtures. When incorporating WMR material, such as the crumb rubber used in the present research, rubberized mixtures were less impactful than the corresponding reference SMA, while a more complex trend was observed compared to AC mixtures. In this sense, the mix design is still a critical factor.

3 Achievements of the TC

TC 279-WMR was active for five years. Its outputs consist of the present State of the Art Report, several journal papers and peer-reviewed conference proceedings, the organization of dedicated workshops, and the dissemination and communication of the results in the form of presentations. As a result of the work of the TGs, various recommendations have been made in the dedicated chapters. These recommendations address different aspects of using waste and marginal materials for roads, such as specimen preparation, testing methods, and type of data needed, and will be published in a separate publication. A detailed list of the TC achievements at the date of publication of this document is presented below.

3.1 State-of-the-Art Report

The STAR Report summarizes the activities of the TC and the main conclusions. In addition, the research needs for follow-up RILEM TCs are identified.

3.2 Peer Reviewed Publications

Six journal papers and conference proceedings were prepared and published at the time of the publication of this STAR document [1 6]:

1. Poulikakos LD, Pasquini E, Tušar M, Hernando D, Wang D, Mikhailenko P, Pasetto M, Baliello A, Cannone Falchetto A, Miljković M, Orešković M, Viscione N, Saboo N, Orozco G, Lachance-Tremblay É, Vaillancourt M, Kakar MR, Bueche N, Stoop J, Wouters L, Dalmazzo D, Pinheiro G, Vasconcelos K, Moreno Navarro F (2022). RILEM interlaboratory study on the mechanical properties of asphalt mixtures modified with polyethylene waste. J Cleaner Prod 375: 134,124. https://doi.org/10.1016/j.jclepro.2022.134124
2. Wang D, Baliello A, Poulikakos LD, Vasconcelos K, Kakar MR, Giancontieri G, Pasquini E, Porot L, Tušar M, Riccardi C, Pasetto M, Lo Presti D, Cannone Falchetto A (2022). Rheological properties of asphalt binder modified with waste polyethylene: An interlaboratory research from the RILEM TC WMR. Resour Conserv Recycl 186: 106,564. https://doi.org/10.1016/j.resconrec.2022.106564
3. Tušar, M, Kakar, MR, Poulikakos LD, Pasquini E, Baliello A, Pasetto M, Porot L, Wang D, Cannone Falchetto A, Dalmazzo D, Lo Presti D, Giancontieri G, Varveri A, Veropalumbo R, Viscione N, Vasconcelos K, Carter A (2022). RILEM TC 279 WMR round robin study on waste polyethylene modified bituminous binders: advantages and challenges. Road Mater Pavement Des: 1–29. https://doi.org/10.1080/14680629.2021.2017330

4. Wang D, Baliello, A, Pinheiro, G, Poulikakos LD, Tušar M, Vasconcelos K, Kakar MR, Porot L, Pasquini E, Giancontieri G, Riccardi C, Pasetto M, Lo Presti D, Cannone Falchetto A (2022). Rheological behaviors of waste polyethylene modified asphalt binder: statistical analysis of inter-laboratory testing results. J Test Eval 51(4):1–2. https://doi.org/10.1520/JTE20220313
5. Pasetto M, Baliello A, Pasquini E, Poulikakos LD (2022). Dry Addition of Recycled Waste Polyethylene in Asphalt Mixtures: A Laboratory Study. Materials 15(14): 4739. https://doi.org/10.3390/ma15144739
6. Pasetto M, Pasquini E, Giacomello G, Moreno-Navarro F, Tauste-Martinez R, Cannone Falchetto A, Vaillancourt M, Carter A, Viscione N, Russo F, Skaf M, Orešković M, Cristina Freire A, Mikhailenko P, Poulikakos LD. (2020, December). An interlaboratory test program on the extensive use of waste aggregates in asphalt mixtures: preliminary steps. In RILEM International Symposium on Bituminous Materials (pp. 215–221). Springer, Cham.

3.3 Organization of Workshops and Presentations

Three workshops were planned during the life of the TC. In addition, further dissemination was performed as part of training activities for doctoral students.

- Workshop 1: December 17, 2020, 9:00–12:00 in conjunction with the ISBM Lyon 2020: International Symposium on Bituminous Materials. Seventy participants attended this workshop online;
- Workshop 2: June 15, 2023, at the end of the TC in conjunction with EATA 2023 10th International Conference of the European Asphalt Technology Association (EATA), Gdansk, Poland. Detailed recommendations for enhanced and new test protocols and standards;
- Workshop 3: July 6, 2023, at the end of the TC in conjunction with the 13th International Conference on Road and Airfield Pavement Technology (ICPT 2023), Beijing, China. Detailed recommendations for enhanced and new test protocols and standards;
- Presentations of TC activities at the Mediterranean Ph.D. School (Naples, Italy, 2021) and SIIV Summer School (Naples, Italy, 2022).

4 Research Needs and Perspective

As remarked in the previous section of this chapter, the work performed by the TC 279-WMR is the first attempt at the RILEM level and internationally through interlaboratory activities to investigate the possibility of using waste and secondary materials for roads. The selected approach, commonly adopted in RILEM technical committees, where interlaboratory experimentation is performed, represents a distinct novelty element. The contributions of many international partners provided solid support to the research results while generating a large volume of data enabling the possibility of identifying the diversity in the behaviour of alternative paving materials and the complexity of the research. Each active TG has identified specific challenges and research needs to be addressed in future activities within RILEM and beyond.

- TG1 suggested relying on mixtures rather than on binder blends to evaluate the performance of the PE-modified paving materials while focusing on a balanced mix design addressing the global response over a wide temperature range. In addition, environmental aspects encompassing a cradle-to-grave approach and the diversity associated with geographical location and specific waste management practices should be considered.
- Based on the work performed on crumb rubber from end-of-life tires, TG2 proposed concentrating the next phase of the research on the chemical investigation to advance the knowledge of the interaction between rubber and binder. At the same time, more understanding is needed concerning the dry mixing process of treated and modified crumb rubber.
- The research performed by TG3 has highlighted the need for investigating and developing mixture performance criteria when operating with waste and secondary materials. Moreover, the demand for increasing the Technology Readiness Level (TRL) is strictly linked to several actions, including field testing, dissemination activities, environmental and life cycle assessment of the performance, and dedicated quality control.
- The work of TG4 was limited to a cradle-to-gate assessment. Therefore, it is in the initial and exploratory phase. On the one hand, defining and adopting benchmark values as an approach to discriminating across paving mixtures appeared promising. However, an extensive data set of case studies for traditional and alternative asphalt mixtures should be used to generate these benchmarks. Hence, it is necessary to start developing and adopting standardized data collection processes at a larger scale to facilitate the analysis in the future.

Beyond the work performed by the different TGs, there is a need to identify the potential sources of pollutants in the recycled materials and their release processes (i.e., polyaromatic hydrocarbons (PAH) content and leaching behaviour of the specific materials) and their effect on health, soils, water, and air, to take the necessary precautions to ensure the safe application of recycled materials in road construction.

The open research questions identified by the different Task Groups (TGs) and the recommendations proposed to extend the work of the TC 279-WMR demonstrated the need for continuing the scientific investigation on waste and secondary materials for roads, possibly expanding the research beyond the combination with traditional paving materials, such as asphalt binders, toward an enhanced circularity vision. Therefore, a demand exists to advance the work of the TC 279-WMR into follow-up research efforts both in the scientific community and specifically within the RILEM umbrella through a new technical committee investigating alternative paving materials. Particular attention should be devoted to characterisation methods, manufacturing procedures, mix design formulation, performance-engineered testing protocols, recyclability potential, and sustainability assessment. Such a new TC should operate to overcome the conventional definition of asphalt mixtures toward the idea of paving mixtures in which the different components (aggregate, binder, additive, and admixtures) will be addressed independently of their origin and source, and next ranked based on their laboratory and field performance as a composite material. This approach would eliminate the discrimination between conventional, secondary, and alternative materials.

References

1. Poulikakos LD, Pasquini E, Tušar M, Hernando D, Wang D, Mikhailenko P, Pasetto M, Baliello A, Cannone Falchetto A, Miljković M, Orešković M, Viscione N, Saboo N, Orozco G, Lachance-Tremblay É, Vaillancourt M, Kakar MR, Bueche N, Stoop J, Wouters L, Dalmazzo D, Pinheiro G, Vasconcelos K, Moreno Navarro F (2022) RILEM interlaboratory study on the mechanical properties of asphalt mixtures modified with polyethylene waste. J Cleaner Prod 375:134124. https://doi.org/10.1016/j.jclepro.2022.134124
2. Wang D, Baliello A, Poulikakos LD, Vasconcelos K, Kakar MR, Giancontieri G, Pasquini E, Porot L, Tušar M, Riccardi C, Pasetto M, Lo Presti D, Cannone Falchetto A (2022) Rheological properties of asphalt binder modified with waste polyethylene: an interlaboratory research from the RILEM TC WMR. Resour Conserv Recycl 186:106564. https://doi.org/10.1016/j.resconrec.2022.106564
3. Tušar, M, Kakar, MR, Poulikakos LD, Pasquini E, Baliello A, Pasetto M, Porot L, Wang D, Cannone Falchetto A, Dalmazzo D, Lo Presti D, Giancontieri G, Varveri A, Veropalumbo R, Viscione N, Vasconcelos K, Carter A (2022) RILEM TC 279 WMR round robin study on waste polyethylene modified bituminous binders: advantages and challenges. Road Mater Pavement Des 1–29. https://doi.org/10.1080/14680629.2021.2017330
4. Wang D, Baliello, A, Pinheiro, G, Poulikakos LD, Tušar M, Vasconcelos K, Kakar MR, Porot L, Pasquini E, Giancontieri G, Riccardi C, Pasetto M, Lo Presti D, Cannone Falchetto A (2022) Rheological behaviors of waste polyethylene modified asphalt binder: statistical analysis of inter-laboratory testing results. J Test Eval 51(4):1–2. https://doi.org/10.1520/JTE20220313
5. Pasetto M, Baliello A, Pasquini E, Poulikakos LD (2022) Dry addition of recycled waste polyethylene in asphalt mixtures: a laboratory study. Materials 15(14):4739. https://doi.org/10.3390/ma15144739
6. Pasetto M, Pasquini E, Giacomello G, Moreno-Navarro F, Tauste-Martinez R, Cannone Falchetto A, Vaillancourt M, Carter A, Viscione N, Russo F, Skaf M, Orešković M, Cristina Freire A, Mikhailenko P, Poulikakos LD (2020) An interlaboratory test program on the extensive use of waste aggregates in asphalt mixtures: preliminary steps. In: RILEM international symposium on bituminous materials. Springer, Cham, pp 215–221

The manufacturer's authorised representative in the EU is Springer Nature Customer Service Centre GmbH, Europaplatz 3, 69115 Heidelberg, Germany. If you have any concerns regarding our products, please contact ProductSafety@springernature.com

Printed and bound by CPI Group (UK) Ltd, Croydon, CR0 4YY
01/10/2026
02225919-0008